Mastering Essential English Skills

Viviane McClelland
Supervisor of Language Development
Richmond, British Columbia

Jerry D. Reynolds
English Consultant
Rochester Public Schools
Rochester, Minnesota

Marion L. Steet
Principal
Penn Treaty Junior High School
Philadelphia, Pennsylvania

Ivory Guillory
English Resource Teacher
St. Landry Parish School Board
Opelousas, Louisiana

Doubleday Canada Limited
Toronto
Doubleday Australia
Sydney

Project Director: Ralph W. Rausch

Staff Editor: Beryl Dwight

Associate Editor: Janet Searcy

Production Director: LaVergne Niequist

Production Assistant: Mary J. Hennessey

Art Director: Gloria Muczynski

Cover Designer: Tak Murakami

Artists: Tak Murakami, Paul Hazelrigg

Printed in Canada

CONTENTS

Using the Right Words

Mastering the Mechanics

Improving Spelling Skills

Building Vocabulary

TO THE STUDENT

Your Two Englishes You're not alone. Everyone has had problems using the English language as effectively as it should be used.

Many, if not most, of your problems with English develop when you forget that there are *two* closely related but essentially different kinds of English—spoken English and written English. To use the language effectively, you must be able to switch from one of its forms to the other with ease. If these two forms of English were identical, you could simply apply one set of rules to both, and many of your problems would disappear. But, unfortunately, spoken English and written English are not the same thing. And you simply can't ignore their differences.

When you speak, you don't have to worry about spelling, punctuation and capitalization, or neatness and legibility. But when you write, these things become very important. When you speak, you can correct yourself immediately if your listener doesn't understand. But when you write, your writing must stand alone and explain itself without you. When you speak, your words vanish in the air. But when you write, they remain for everyone to see. Small wonder that speaking seems so easy and natural; writing, so difficult and forced. Small wonder, too, that others are more critical of the way you write than they are of the way you speak.

Rules for "Good" English Because people from different parts of the country and from different backgrounds speak English differently, it's very difficult, if not impossible, to establish hard-and-fast rules for a standard spoken English. But while people may expect varieties of spoken English to "sound" different, they expect written English to "look" the same. This is why fairly rigid and universal standards for written English have been established and why these standards are taught in schools. In fact, the sort of "good" English an educated person is expected to use is called Standard English—or, more accurately, Standard Written English.

The purpose of this book is to help you increase your skill in using Standard Written English—and, in the process, to help you become better acquainted with the rules that govern "good" English.

Throughout the book you'll find some example sentences and expressions identified as incorrect. In this book *incorrect* means that while these sentences and expressions may be acceptable in casual conversation, they will probably not be acceptable in more formal speaking situations. They will certainly not be acceptable in your writing. The word *formal* is used to identify writing and speaking situations in which you'll have to be very careful to follow the rules of Standard Written English. The word *informal* is used to indicate those situations in which you need not be so careful in following the rules. Generally, informal situations are those in which you are among friends who are mostly interested in *what* you say or write. Formal situations, however, are those in which people are concerned not only with *what* you say or write but also with *how* you say or write it. Most of the advice about "good" English given in this book is about formal writing situations. But a good deal of it applies to formal speaking situations as well.

Essential English Skills To be successful in school and in the workaday world, you'll have to demonstrate your mastery of the basic skills necessary for using English effectively. These essential English skills include being able to write clear, complete, well-constructed sentences; being able to use the right word at the right time; being able to punctuate and capitalize correctly; being able to spell correctly; and being in command of a good-sized vocabulary.

To help you master these essential English skills, this book presents sixty-five lessons grouped under titles that identify five important skills areas: "Solving Sentence Problems," "Using the Right Words," "Mastering the Mechanics," "Improving Spelling Skills," and "Building Vocabulary." Following your teacher's directions for studying these lessons and then practicing the skills they teach will help you become a better writer—and speaker—of "good" English.

Solving Sentence Problems

1 Sentence Fragments

Sentence fragments are parts of sentences. Very often they are phrases and clauses that have been incorrectly written as complete sentences. You can correct a phrase or clause fragment either by joining it to a complete sentence or by rewriting it to make it a complete sentence. Some kinds of fragments are best eliminated by joining. Other kinds can be corrected either by joining or by rewriting. What follows are six typical kinds of sentence fragments with suggestions for correcting them.

practice 1

A prepositional phrase is a group of words introduced by a preposition. Examples: ***on** the road, **in** the cabinet, **to** everyone, **before** the intermission.* Prepositional phrases are parts of sentences. They should not be written as complete sentences.

We walked home in the rain. ***After the inter-high track meet.**

We walked home in the rain after the inter-high track meet.

Correct the sentence fragment in each of the following items by joining it to the sentence. The fragment is in parentheses.

1. Dad finally found a parking space. (Near the big new supermarket.)
2. The ice on the river has melted slightly. (Under the bridge near the old mill.)
3. Only our teacher understands the process. (From beginning to end.)
4. Gymnasts lose points if their feet hit the floor. (During their performance on the uneven parallel bars.)
5. Storm warnings are up along the coast. (From Glace Bay to Yarmouth.)
6. Turn left when you reach the intersection. (By the old railway station.)
7. Maureen left her algebra book. (Beside the bicycle rack at the main entrance.)

*The asterisk (*) identifies incorrect sentences.

8. When we were little, Mom used to hide the cookie jar. (Behind the cereal boxes.)
9. Mr. Russo said that the gym will be closed next week. (For repairs and painting.)
10. Nobody laughed at the softball team. (After their victory last Saturday.)

practice 2

An appositive is a word or group of words that restates, identifies, or explains another word. Examples: *his dog, **the beagle**; my oldest sister, **Margot**; the first prize, **a ten-speed bike.*** Appositives are parts of sentences. They should not be written as complete sentences.

> Marty has just met Mr. Berman. ***The new swimming coach from Ontario.**
>
> Marty has just met Mr. Berman, the new swimming coach from Ontario.

Correct the sentence fragment in each of the following items by joining it to the sentence. Use a comma before the fragment.

1. Dave's new friend is Rover. A gigantic two-year-old Saint Bernard.
2. Tonight we play Lawrence Park Collegiate. Last year's provincial basketball champions.
3. My biology report will be on photosynthesis. The formation of carbohydrates by plants exposed to light.
4. Among the election winners was Simma Holt. Well-known Vancouver journalist.
5. The ingenious rotary engine is named for its German inventor. Felix Wankel.
6. "Mood Indigo" was written by Duke Ellington. One of North America's most influential jazz musicians.
7. Probably the best-known cat in North America is Morris. A television star for cat-food commercials.
8. The sixth-grade students sent get-well cards to Skana. The sick whale at the aquarium.
9. Beth ordered blueberry pancakes with syrup. Her favorite breakfast.
10. That trailer belongs to the Chandramohans. The family next door who just moved in last week.

practice 3

Participial, gerund, and infinitive phrases all contain verb forms that are not the main verb of a sentence. Examples: ***running*** *down the street, after* ***watching*** *the game,* ***to play*** *tennis well.* Such phrases are parts of sentences. They should not be written as complete sentences.

Pete worked all evening. ***Finishing his social studies report at twelve.**

Pete worked all evening, finishing his social studies report at twelve.

Pete worked all evening. **He** finished his social studies report at twelve.

Laura's responsibilities are many. ***From drawing up contracts to issuing public statements.**

Laura's responsibilities are many, from drawing up contracts to issuing public statements.

Laura's responsibilities are many. **They range** from drawing up contracts to issuing public statements.

We looked through old newspapers. ***To find the article about Marie's walking the Cabot Trail.**

We looked through old newspapers to find the article about Marie's walking the Cabot Trail.

Correct the sentence fragment in each of the following items in two ways. First, join the fragment to the sentence that precedes it. Use a comma before the fragment in items 1, 4, 5, 9, and 10. Second, make the fragment a complete sentence.

1. Tom walked home slowly. Discouraged after failing the exam.
2. On weekends we earned money for the trip. By washing cars and baby-sitting.
3. Paula usually feeds the dogs. Before leaving for school.
4. My brother unintentionally ate some yogurt. Thinking it was ice cream.
5. I didn't answer. Not having understood the question.
6. Kevin couldn't find the path. Covered by the drifting snow.
7. Carol decided to see the movie. After hearing us talk about it.

8. Elaine Chang wants a chance to practise more. Before trying out for the team.
9. Mike drove right onto the unpaved road. Not noticing the detour sign.
10. One of the newest winter sports to gain worldwide attention is freestyle skiing. Commonly called "hotdogging."

Correct the sentence fragment in each of the following items by joining the fragment to the sentence.

11. We all stood around Ms. Hansen's desk. To watch the experiment.
12. Diane waited until five. To get tickets for the basketball game.
13. Everyone tried to encourage Rosie. To try out for the team.
14. Hundreds of people waited along Main Street. To see the mayor and her family.
15. On the late show, the mad scientist worked frantically. To complete his monster before dawn.

practice 4

A compound verb consists of two verbs joined by a connecting word like *and, but, or.* Examples: ***tripped*** *and* ***stumbled, is failing*** *but* ***is continuing, go swimming*** *or* ***play ball.*** Both parts of a compound verb belong together in the same sentence. The connecting word and the second verb should not be written as a complete sentence.

> Tom looked all over the house for his gloves. ***And finally found them in his pocket.**
>
> Tom looked all over the house for his gloves and finally found them in his pocket.
>
> Tom looked all over the house for his gloves. **He** finally found them in his pocket.

Correct the fragment in each of the following items in two ways. First, join the fragment to the sentence that precedes it. Second, make the fragment into a complete sentence.

1. Joe and Carl entered the three-legged race. And won it.
2. In auto shop we flushed out the radiator. But forgot to replace the radiator cap.

3. Our club president asked for quiet. And called the meeting to order.
4. Dr. Norman Bethune has long been a hero in China. But has only recently become well-known in his native Canada.
5. Zarin is on the stage crew. And also plays in the band.
6. During the storm the bus skidded on the wet pavement. And slid into the ditch.
7. Raoul is good at biology. But has a hard time with math.
8. The ground crew dragged the infield. And put down the bases.
9. As she ran for the bus, Diane grabbed her books. But left her lunch on the table.
10. The guest speaker talked for 30 minutes. And then answered questions from the audience.

practice 5

Adjective clauses usually begin with a connecting word like *who, whom, whose, which,* or *that.* They are not questions. Examples: ***who** was late,* ***which** was broken,* ***that** we noticed.* Adjective clauses are parts of sentences. They should not be written as complete sentences.

> Pat came with a key. ***Which didn't fit any of the locks.**
>
> Pat came with a key which didn't fit any of the locks.
>
> Pat came with a key. **It** didn't fit any of the locks.

Correct the fragment in each of the following items in two ways. First, join the fragment to the sentence. Use a comma before the fragment in items 3, 5, 7, and 9. Second, make the fragment into a complete sentence.

1. Dad is going to test-drive the motorcycle. That Russ wants to buy.
2. Joanne interviewed the student. Whose photograph won the prize.
3. In Canada, interest has been growing in soccer. Which is said to be the world's most popular sport.
4. Nobody understands the solution. That Karen put on the board.
5. Several best-selling record albums are by Paul Anka. Who was born in Ottawa in 1941.

6. I sat next to the basketball player. Whom we met last night.
7. When the Soviet gymnastics team appeared at the 1976 Olympics in Montreal, everyone wanted to see Olga Korbut. Who had three Olympic gold medals at the age of 17.
8. Try to avoid loud noises. Which will wake the baby.
9. Tomorrow my sister and I have to clean out the garage. Which has been collecting odds and ends all winter.
10. The salesperson wouldn't wait on the customer. Who had pushed to the head of the line.

practice 6

Adverb clauses usually begin with a connecting word like *because, although, when,* or *if.* Examples: ***because** I want to, **if** you say so, **although** it rained.* Adverb clauses are parts of sentences. They should not be written as complete sentences.

> Pierre takes his camera with him. *Wherever he goes.
> Pierre takes his camera with him wherever he goes.

Correct the fragment in each of the following items by joining the fragment to the sentence.

1. You can repair this toaster yourself. If you want to.
2. Michael slipped into his seat. Just as the band began to play.
3. Jo's aunt and cousins were waiting for her. When she arrived at the bus terminal.
4. Al explained the solution again. Since nobody understood it the first time.
5. We decided to wear ponchos and boots on the hike. After we heard the weather forecast.
6. Linda plays volleyball three afternoons a week. Because she enjoys it.
7. My brother started taking guitar lessons last month, and now he practices faithfully. Whenever he has a chance.
8. Suzy didn't know the answers. Because she hadn't read the chapter.
9. Tony decided to apply for a summer job with the park department. Although he knew that the competition would be stiff.
10. I always confused the crankshaft and the camshaft. Until I saw a model of a working piston engine.

application

Rewrite the following paragraphs, correcting each fragment. Make the fragment a complete sentence or join it to a sentence, whichever seems more effective.

> No one has yet solved the mystery of the Bermuda Triangle. An area of ocean off the coast of Florida. Where over fifty ships and planes have disappeared. In most cases without leaving a trace. The area covers about 440,000 square miles. From Florida to Bermuda to Puerto Rico and back to Florida.
>
> One of the strangest incidents occurred in 1945. When five U.S. bombers disappeared on a training mission. Nobody expecting trouble. But a seaplane also vanished when it took off. Carrying an experienced crew. To search for the five bombers.
>
> There is no good explanation of these events. Most of which occurred in good weather. Scientists think that perhaps unexpected storms or downward air currents cause the accidents. And speculate that ocean currents may carry the wreckage away.

2 Run-on Sentences

You write a run-on sentence whenever you join sentences without putting a punctuation mark or a conjunction between them.

*The kindling was damp the fire would not catch.

You also write a run-on sentence whenever you join sentences with only a comma between them.

*The kindling was damp, the fire would not catch.

This kind of run-on sentence is also called a comma fault or comma splice.

Here are four ways to correct run-on sentences:

1. Write each of the sentences as a separate sentence. Use end punctuation after the first sentence, and begin the second with a capital letter.

The kindling was damp. The fire would not catch.

*The asterisk (*) identifies incorrect sentences.

2. Join the sentences together using a semicolon between them.

The kindling was damp; the fire would not catch.

3. Join the sentences together using a comma and a conjunction like *and, but, or* between them.

The kindling was damp, **and** the fire would not catch.

4. Rewrite one of the sentences making it a subordinate clause or phrase.

Because the kindling was damp, the fire would not catch.

The kindling being damp, the fire would not catch.

The kindling was **so** damp **that the fire would not catch.**

Watch out for run-on sentences whenever you use words like *therefore, however, nevertheless,* and *furthermore.* These words can cause problems because they may be used as either interrupters or connectives. When you use them as interrupters in the middle of a sentence, you set them off with commas.

The kindling was damp. The fire, **therefore,** would not catch. [interrupter]

But when you use such words as connectives, you must use a semicolon between the sentences they connect. Or, you must end the first sentence with a period and begin the connective with a capital letter.

The kindling was damp; **therefore,** the fire would not catch. [connective]

The kindling was damp. **Therefore,** the fire would not catch. [connective]

practice 1

If the following sentences have been correctly joined, write *correct* after the number. If not, write *run-on* after the number.

1. The jury voted to acquit the defendant was jubilant.
2. Sally and I swapped hockey cards, I gave her a Ken Dryden for a Bobby Orr.

3. Lynn plays squash twice a week at the Y, she runs three miles every morning.
4. The wind shifted, but the smog remained.
5. Although the bike was secondhand, it was in good condition.
6. The plane broke through the cloud cover I could see the airport.
7. Genya Ravan is her stage name; her real name is Goldie Zelkowitz.
8. In road tests the car averaged twenty-seven miles per gallon, in everyday use it averages twenty-one.
9. The morning passed quickly however the afternoon dragged on—and on.
10. The newsstand on the corner sells *Maclean's* and *Saturday Night,* it also sells *Chatelaine* and *Canadian Motorist.*

practice 2

The following sentences are all run-on sentences. Rewrite each one, adding end punctuation and capitalization to make two separate sentences.

1. Backpacking is fun in most provincial parks foot trails have been marked out.
2. June is the best month for fishing, July is usually the worst.
3. A bale of hay weighs 300 pounds a bale of cotton weighs 400 pounds.
4. Roll an orange vigorously on a hard surface before squeezing it, you'll get a lot more juice.
5. Do you know Jill Hatfield's remedy for hiccups it's to eat a spoonful of peanut butter.
6. Stephen Leacock was a Canadian writer who achieved a world-wide reputation, one of his best known stories is entitled "My Financial Career."
7. The wind's in our face, and the sun's at our back, we can photograph the deer safely from here.
8. Arnie took a Stretch & Sew class now he can make his own suits and sport coats.
9. The climbers bedded down on a narrow ledge they roped themselves to the mountain wall so they wouldn't roll off.
10. Geologists are often employed by oil companies they use aerial photographs to find oil deposits.

practice 3

The following sentences are all run-on sentences. Rewrite each one, adding a semicolon.

1. There was to have been an eclipse of the moon last night an astronaut on the moon would have seen an eclipse of the Earth.
2. Wanda has a new 35 mm reflex camera she's working on the *Daily Times-Press* this summer.
3. I bought a recording of Scott Joplin's rags they're played by Joshua Rifkin.
4. Track events involve running field events involve jumping and throwing.
5. Baking soda makes an effective remedy for insect bites mixed with a little water it can be pasted right on the bite.
6. Art copies reality it's a mirror that reflects nature.
7. The engine was sound, however, it badly needed a tune-up.
8. *4'33"* by John Cage is a musical composition it consists entirely of four minutes thirty-three seconds of silence.
9. Work when it's time to work play when it's time to play.
10. His den was full of handsomely bound books the trouble was he never read any of them.

practice 4

The following sentences are all run-on sentences. Rewrite each one, adding a comma and an appropriate conjunction.

1. No member finished in the top five the team won the meet.
2. Slogans are not solutions a catchy phrase is not insight.
3. We have a legal responsibility to obey the law we have a moral responsibility to question it.
4. Wrench the bolt firmly don't force it.
5. A baseball glove needs to be oiled regularly the leather will crack.
6. To be wrong is a fault it's worse to be stubborn.
7. The hem ought to be double stitched it might fall out.
8. Intelligence is a good thing honesty is better.
9. The four of us climbed into the Land Rover we were soon headed up a switchback into the mountains.

10. Fortunately we caught some trout we would have lunched on biscuits.

practice 5

The following sentences are all run-on sentences. In each run-on, rewrite one of the sentences to make it a subordinate clause or phrase.

1. The undertow below the dam is strong we shouldn't swim there.
2. Marigolds keep insects out of gardens they should be planted along the borders.
3. A three-speed bike is more practical than a five-speed, the two extra speeds are of little value.
4. The bricks aren't here we can't lay the patio.
5. You're going sailing remember to wear a life jacket.
6. Margaret Laurence is a well-known Canadian writer, her novel *A Jest of God* was made into the movie *Rachel, Rachel.*
7. I was jogging through the forest preserve, I heard a wood thrush singing.
8. Lester B. Pearson in 1957 won a Nobel Peace Prize, he became Prime Minister of Canada in 1963.
9. You enter the city, the smell from the refineries and mills hits you first.
10. The CN Tower in Toronto is the world's tallest self-supporting structure, it is 556 metres high.

← application →

The following paragraph has been incorrectly written as one long run-on sentence. Rewrite the paragraph, breaking it into several sentences. See if you can apply each of the following methods at least once: (1) add end punctuation and a capital letter; (2) add a semicolon; (3) add a comma with a conjunction; (4) rewrite some sentences, turning them into subordinate clauses or phrases.

> One purpose of education is to prepare students for the real world the American state of Oregon has adopted some new statewide graduation requirements intended to do just that standard academic require-

ments will continue by 1978 in order to graduate from high school in Oregon, you'll have to be able to accomplish a variety of practical, everyday tasks, for example, you'll have to be able to balance a checkbook, fill out a 1040 income-tax form, demonstrate a knowledge of basic first aid, even plan a balanced diet, in addition, individual school districts have added their own requirements in Corvallis, Oregon, students must be able to officiate two different sports in Eugene they must be able to read a freeway road map in Philomath they must explain the purpose of various paycheck deductions some parents think these requirements are too tough the students themselves are unconcerned many people in Canada would like to see provincial departments of education introduce similar measures.

3 Choppy Sentences

When you write, put closely related ideas into the same sentence. Don't write them as separate sentences. When closely related ideas are kept apart, the result is usually a string of short, choppy sentences like this:

> The classroom [1] was crowded. Mr. Guthrie came [2] into the classroom. Everyone in the [3] classroom stopped talking. Mr. Guthrie could [4] not figure out why.

Notice that these four sentences contain needless repetitions. For example, *Mr. Guthrie* and *the classroom* are repeated several times. Notice, too, that a minor detail like the crowded condition of the classroom is given a complete sentence of its own. And because the related ideas are expressed as separate sentences, their relationship is unclear.

You can avoid this choppy style by combining several sentences into one. For example, see how needless repetition disappears and related ideas are brought together when you combine the four sentences:

Mr. Guthrie could not figure out why everyone stopped talking when he came into the crowded classroom.

There are three ways to combine sentences effectively. Whichever way you choose will require some rewriting of the original sentences.

1. Combine equally important sentences when one sentence adds to or contrasts with the other. To do this, join the sentences with a comma and a coordinating conjunction like *and, but, or, nor, for, so, yet;* with a semicolon and a connective like *however, therefore, nevertheless, moreover, furthermore;* or with a semicolon alone.

> The performance was supposed to begin at eight o'clock. Some of the costumes were lost at the airport.
>
> The performance was supposed to begin at eight o'clock, **but** some of the costumes were lost at the airport.

> The gas tank was full, the oil and coolant were checked, and the tires were properly inflated. I was wary about the drive across the desert.
>
> The gas tank was full, the oil and coolant were checked, and the tires were properly inflated; **nevertheless** I was wary about the drive across the desert.

2. Combine sentences of unequal importance when the less important sentence merely supplies a detail of time (*when*), condition (*if*), or identification (*who*). To do this, change the less important sentence into a subordinate clause beginning with a subordinating conjunction like *after, although, as, because, before, if, since, unless, when, while.* Or change the less important sentence into a subordinate clause beginning with a relative pronoun like *who, which,* or *that.* Let the more important sentence stand as the main clause.

> **Try to start the engine.** Press the gas pedal down once or twice.
>
> **Before you try to start the engine,** press the gas pedal down once or twice. [time]

The friedcakes glistened with beads of sugar crystals. **The friedcakes were fresh and still warm.**

The friedcakes**, which were fresh and still warm,** glistened with beads of sugar crystals. [identification]

3. Combine sentences of unequal importance when the sentences contain closely related details of description or explanation. To do this, reduce the sentences of less importance to phrases or to single words. Then insert them into the more important sentence to act as modifiers, appositives, or other sentence parts.

The media class took over a radio station for one afternoon. **That station was CJOR.**

The media class took over a radio station**, CJOR,** for one afternoon. [appositive]

Snow fell throughout the day. **This snowfall was blinding. It was accompanied by strong winds.**

Blinding snowfall**, accompanied by strong winds,** fell throughout the day. [modifiers]

practice 1

Rewrite each of the following pairs of sentences as a single sentence. Combine each pair by using the elements indicated in parentheses.

1. The band played a concert early in the evening. The fireworks went off after dusk. (*comma and coordinating conjunction*)
2. Kathleen nominated Maria. Maria declined the nomination. (*semicolon and connective*)
3. The prophet forecast impending doom. No one paid him any attention. (*comma and coordinating conjunction*)
4. Popular music is one thing. Acid rock is quite another. (*semicolon*)
5. The tapestry was beautiful. It was too expensive. (*comma and coordinating conjunction*)
6. Her income is modest. She enjoys teaching. (*comma and coordinating conjunction*)
7. The class play should be advertised. Attendance will be disappointing. (*comma and coordinating conjunction*)

8. Grandma looks younger than seventy-five. She acts younger. (*semicolon and connective*)
9. The sunrise over the lake was spectacular. It was worth getting up for. (*semicolon*)
10. I have a super pattern for a dress. I know just the material to use. (*comma and coordinating conjunction*)

practice 2

Rewrite each of the following pairs of sentences as a single sentence. Combine each pair by changing one of the sentences into a subordinate clause beginning with a subordinating conjunction. Or, if necessary, change one of the sentences into a subordinate clause beginning with a relative pronoun.

1. The steak was charred on the outside. Inside it was pink.
2. The puck skipped past the goalie into the net. The puck was bouncing crazily.
3. Dad was listening to a baseball game on the radio. He was watching a tennis match on TV.
4. The government's Anti-Inflation Act has not been popular. It attempts to control wages and prices.
5. *McCabe and Mrs. Miller* was directed by Robert Altman. The film was shot in British Columbia.
6. The wind died down. The boat was in danger of capsizing.
7. We are visiting San Francisco. One of our chief delights is riding the cable cars.
8. Karen Kain and Frank Augustyn won a nine minute standing ovation when they appeared with the Bolshoi Ballet in Moscow. They are the first Canadians to appear with this world famous company.
9. We walked along the beach. We returned to our camper.
10. The weather conditions were threatening. A tornado watch was issued in southern Manitoba.

practice 3

Rewrite each of the following groups of sentences as a single sentence. Combine them by reducing one or two of the sentences in each group to a phrase or to a single word. Then insert the phrase or word into the remaining sentence as a modifier, an appositive, or other sentence part.

1. Alligators are a menace in certain populated areas of Florida. They were once almost extinct.
2. The speaker was the member from Burnaby North. The member's name is Eileen Dailly.
3. Thousands of spectators came. They came by car. They came by bicycle. They came on foot.
4. We visited the McMichael Canadian Collection. The collection is in the village of Kleinburg, Ontario. It includes some 900 works by Canadian artists.
5. Croquet is a lawn game. It is a polite, leisurely game. It is a game sometimes played with surprising vengeance.
6. Multicolored pennants flapped in the breeze. The pennants were made of vinyl. The breeze was stiff.
7. I leaned against the fence. I watched a horse. The horse was beautiful. The horse was grazing. The horse was in the lower meadow.
8. The storm came suddenly. It blew in from the west. It brought hailstones. The hailstones were the size of golf balls.
9. The books were schoolbooks. The books were piled in stacks. The stacks were crooked. The stacks were in the corner of the library.
10. The game was over. The game was for the championship. The championship was of the city. The city was Montreal.

← application →

The following paragraphs have been written entirely as a string of short, choppy sentences. Rewrite the paragraphs, combining sentences wherever appropriate. Use as many different methods of combining as you can. Read your paragraphs aloud to see if they flow smoothly.

> An automobile accident is unavoidable. You can take some manoeuvres. These manoeuvres will lessen car damage. They will lessen personal injuries. For example, take a head-on crash. To avoid hitting head-on, make a turn. The turn should be sharp. The

turn should be to the right. Then the oncoming car will hit you in the side. The oncoming car will not hit you head-on. You might lose control of your car. Don't worry. Anything is better than this. This is a head-on crash!

Here's another example. Take a rear-ender. Your car is going to be hit. The hit is from behind. Take your foot off the brake. The impact occurs. Slam on your brake immediately. Keep a firm grip on the steering wheel. The firm grip lets your arms absorb some of the shock. The shock is from the impact. Slouch down. Your head is below the back of the seat. The back of the seat will give you protection. This protection will be against whiplash.

4 Faulty Coordination

You use coordination whenever you connect words, phrases, or clauses with a coordinating conjunction like *and, but, or, nor, for, so, yet.*

> **Carla** and **I** both had an exhibit at the science fair. [words]

> Would you prefer **to attend a concert** or **to go shopping?** [phrases]

> **The cabinet members were seated,** but **the Prime Minister was still on the phone with the ambassador.** [main clauses]

You also use coordination whenever you join main clauses with a semicolon and a connective like *however, therefore, nevertheless, moreover, furthermore.*

> **The forecast was for clear weather**; however, **the barometer took a sudden dip.** [main clauses]

Coordination is an effective way to connect related elements in a sentence. However, if the elements you connect have no logical relationship to each other, the result is faulty coordination.

> ****The movie was terrible**, and **we stopped for pizza afterward.**

*The asterisk (*) identifies incorrect sentences.

You can avoid this sort of faulty coordination by making each clause a separate sentence. Then expand the ideas introduced by these two sentences into two separate paragraphs.

> **The movie was terrible.** The plot was unbelievable, and the actors were miscast. Often the music on the sound track was barely audible. . . .
>
> **We stopped for pizza afterward** and met some friends. Even Ellen Portner, who loves musical comedy, was disappointed. . . .

Faulty coordination also results when the clauses you connect are logically related but are of unequal importance.

> *__The football team left the huddle__, but **the school mascot, a goat, ran out onto the field.**

Connecting these two clauses with *but*—that is, coordinating them—indicates that the ideas they express are equally important. However, the really important idea is that the goat ran onto the field. That the team left the huddle is only incidental information that tells when the goat ran. To avoid this kind of faulty coordination, put the less important idea in the form of a subordinate clause beginning with *as, if, when, since, while, because, after, before,* or *although*.

> **As the football team left the huddle,** the school mascot, a goat, ran out onto the field.

In the following sentence, *and* is used to connect, or coordinate, two clauses that are related but are not equally important:

> **Joe Simonsen won the provincial lottery**, and **he is a cattle rancher.**

The fact that Mr. Simonsen is a cattle rancher is less important than the fact that he won the provincial lottery. In sentences like this one, you can avoid faulty coordination by putting the less important idea in the form of a subordinate clause beginning with *who, which,* or *that*.

> Joe Simonsen, **who is a cattle rancher,** won the provincial lottery.

Finally, be careful not to overuse coordination. Don't, for example, write a long rambling series of clauses connected by *and's*.

*I bought a new bike, and **it's a Raleigh**, and **it's a ten-speed**, and **it has hand brakes**, and **the derailleur is by Wockenfuss.**

To avoid such faulty coordination, change some of the clauses into phrases or into single words. If necessary, break the long sentence into two shorter sentences.

I bought a new **ten-speed Raleigh** bike **with hand brakes. The derailleur is by Wockenfuss.**

practice 1

If the connected elements in each of the following sentences are logically related, write *related* after the number. If the connected elements are not logically related, write *unrelated* to indicate faulty coordination.

1. A Plymouth Fury won the race, but the principal drives a Volkswagen.
2. The sun was shining, and Friday was Mother's birthday.
3. We have apples but no oranges.
4. The game is played as scheduled, or season tickets can be purchased at the box office.
5. The heat wave continued, and water rationing became a reality.
6. The reunion took place Sunday afternoon; nevertheless, the sun rose at 5:49 that morning.
7. Central City was the fictional name of the town in the movie; however, several of the scenes were shot on location in Toronto.
8. I prefer country and western music, yet Gordon Lightfoot is appearing at the Coliseum.
9. The tournament was contested over the rolling hills of Pine River Golf Club, but the concession stands were charging $1.00 for a hot dog.
10. Dissent is not disloyalty nor is accusation proof.

practice 2

Rewrite the following sentences, putting the less important idea in the form of a subordinate clause. Begin the subordinate clause with *as, if, when, since, while, before, although, because,* or *after.*

1. The Member of Parliament rose to speak, and the TV networks took a station break.
2. I've always liked working outdoors, and I wish I lived on a farm.
3. The runners were still out on the course, but the thunderstorm struck.
4. He had to practice his cello, but he could not go to the ceremony.
5. Public transit became popular, and gas prices went up.
6. The story begins, and Little Red Riding Hood is preparing a basket of goodies for her grandmother.
7. The first light of dawn brightened the sky, and Angie and Nick were at the pier ready to go fishing.
8. The train reached the head of the curve, and the engineer spoke to the stationmaster by walkie-talkie.
9. Lars knows nothing about mechanics, but I fixed the pump.
10. Farmers' incomes have fallen slightly in this province, and the farmers are expected to vote heavily for the government's reelection.

practice 3

Rewrite the following sentences, putting the less important idea in the form of a relative clause. Begin each relative clause with *who, which,* or *that,* whichever is appropriate.

1. Toller Cranston won the Canadian Men's National Figure Skating Championship six times, and he now has his own ice show.
2. A new wing has been added to the gym, and this wing will house handball and squash courts.
3. Neolithic people were prehistoric farmers, and they made tools out of flint and bone.
4. The book lay on the coffee table, and it was bound in genuine leather.
5. The mercury stood at 5 below, and that's not bad for January in Yellowknife.
6. We recently pooled our savings to purchase a Cessna T-207, and it has a cruising speed of 176 miles per hour.

7. *I Never Promised You a Rose Garden* is a sensitive novel about mental illness, and it was written by Hannah Green.
8. The coach disagreed, and she believes pep pills impair an athlete's performance.
9. Ms. Gradowski has a pilot's licence, and she is my math teacher.
10. Oona's Mustang needs a new fuel pump, but it is up for sale.

practice 4

Rewrite the overcoordinated sentences that follow, changing clauses into phrases or into single words. Where appropriate, break the overcoordinated sentence into two sentences.

1. The committee investigated the agency, and it published a report, but the report was suppressed, and that was unfortunate.
2. The meal consisted of beans and onions, and there was also rice, and coffee and warm milk were mixed and served afterward.
3. I gave a birthday party, and we pinned the tail on the donkey, and we fished for prizes, and we had ice-cream sundaes to eat and cupcakes, too.
4. These shoes are Converse All-Stars, and they have suede leather uppers, but they are basketball shoes.
5. Mother worked in the garden, and she pulled weeds, and she thinned the beets, and she dusted the tomato plants.
6. The doctor inserted one end of a long tube into my throat, and she attached the other end of the tube to a machine, and that machine was a vacuum pump.
7. The columns still stood, but they were gray, and they were flecked with moss.
8. It was a quiet night, and I sat alone at the kitchen table, and through the window I could see fireflies winking at each other.
9. The Little Theatre rehearsed a new play, and it was a "who-dun-it," and it was by Agatha Christie, and auditions were open to the public.

10. I have some new tennis balls, but they are unpressurized ones, and we are playing on clay, so we should use pressurized balls.

application

Read the following paragraphs. The numbers in color indicate where sentences have been taken out.

> LeRoi Anderson was playing in his first golf tournament. **(1)** As Anderson approached the number one tee, there were perhaps fifty spectators waiting idly. Now to those accustomed to jammed freeways and packed beaches, fifty people standing around a golf course might seem like a small crowd indeed. To Anderson, however, fifty people were altogether too many—about forty-eight or forty-nine too many. **(2)**
>
> Anderson bent stiffly to place a golf ball on a tee. As he addressed the ball, he waggled his club awkwardly and peeked out at the fairway winding through the trees.
>
> **(3)** There was a shocked gasp from the crowd, a nervous shuffling of feet, then silence. **(4)** The silence that followed this gaff was oppressive. Sweating profusely, his upper lip twitching, Anderson prepared to attack the ball yet a third time. Just then a voice broke the embarrassed silence:
>
> "Tough course, ain't it, Mac?"
>
> A few hesitant giggles flowed quickly into a burst of laughter. **(5)** Then with the tension visibly broken, he took up his stance, swung smoothly, and stroked the reluctant ball out into the center of the fairway.

The sentences that follow were removed from the paragraphs. Each sentence contains faulty coordination. Rewrite each sentence to get rid of the faulty coordination. Make certain your rewritten sentence will fit smoothly into the paragraph. As you rewrite each sentence, read it and the sentences around it aloud to see if it does fit smoothly. If it doesn't, try rewriting it a different way.

1. He was a fair golfer, but he was a nervous individual, and his game frequently fell apart under stress.

2. They stood far back from the tee, but Anderson could feel their breath on his back, and their breath was heavy.
3. One hundred eyes were burning holes in him, and Anderson took a mighty swing, and he missed the ball completely.
4. His teeth were clenched, and Anderson readdressed the ball, and he jerked the club back, and he swung, and again he missed the ball.
5. Anderson stepped back from the ball, and he sighed, but he offered a weak smile.

5 Faulty Subordination

You use subordination whenever you connect two clauses with a subordinating conjunction like *because, since, before, although.* Because the clause that begins with the conjunction usually expresses the idea of lesser importance, it is called the subordinate clause.

> Radio transmission was interrupted [main clause] **because a power tube blew out** [subordinate clause].

If you want the subordinate clause to express time, begin it with *after, as, before, when,* or *while.* If you want the subordinate clause to express a condition under which something happens, begin it with *although, because, since,* or *unless.*

> The boutique closed **after** the dresses were sold out. [time]
>
> **Unless** the rain stops, the picnic will be held under the pavilion. [condition]

If you want the subordinate clause to describe or identify someone or something, begin it with *who, that,* or *which.*

> The strawberries, **which** we grew in our garden, were large, juicy, and sweet. [identification]

Subordination is an effective way to express two related ideas in a single sentence and to make clear what their relationship is. However, be careful to avoid faulty subordination. For example, if you place the less important idea in the main clause and the more important idea in the subordinate clause, the result is upside-down subordination.

> *The river flooded **when** thousands were forced from their homes.

You can correct this kind of faulty subordination by switching the subordinating conjunction to the beginning of the other clause.

> **When** the river flooded, thousands were forced from their homes.

This switch changes the faulty main clause into an effective subordinate clause. It turns the subordination right side up. Notice that when the subordinate clause comes first in the sentence, a comma is placed between the two clauses.

Faulty subordination also results whenever you place *and* or *but* between a main clause and a subordinate clause beginning with *which* or *who.*

> *The band needs new uniforms for next season **but which** cannot be purchased until the money is raised.
>
> *The school board hired a new English teacher **and who** will also coach track.

To correct this kind of faulty subordination, simply remove the *and* or the *but.*

> The band needs new uniforms for next season **which** cannot be purchased until the money is raised.
>
> The school board hired a new English teacher **who** will also coach track.

Finally, avoid oversubordination, which results when you try to add too many details to your main clause. Don't string together a long series of subordinate clauses referring to some element in a previous clause.

> *Cindy Nicholas ranks as one of the world's great athletes **who** in September 1977 swam the English Channel both ways **where** it separates Dover and Cap Gris Nez **which** is a distance of about 21 miles in a time of 19 hours 55 minutes **that** broke the record set by John Erikson of Chicago in 1975 by more than 10 hours.

*The asterisk (*) identifies incorrect sentences.

To avoid oversubordination, leave out unnecessary details; change clauses into phrases or single words; and, if possible, break up an oversubordinated sentence into two sentences. Note that this may require some rewriting:

> Cindy Nicholas, one of the world's great athletes, swam the English Channel both ways in 1977 in a time of 19 hours 55 minutes. Her 21 mile swim broke the previous record, set by John Erikson, by more than 10 hours.

practice 1

Rewrite the following sentences, turning upside-down subordination right side up. That is, switch the subordinating conjunction to the beginning of the other clause so that the less important idea is in the subordinate clause and the more important idea is in the main clause. Remember to add a comma if the subordinate clause in your sentence comes first.

1. The sun went down after the farmer continued planting by the headlights of the tractor.
2. The first days of school are hectic although old friendships are renewed and new ones started.
3. When we sat down to rest, our legs grew tired.
4. The herd calmed down unless the cowhands would be up all night.
5. Because we can't go surfing today, the water is calm.
6. Although the trophy is not a plate but a gold cup, Canada's most famous thoroughbred race is called the Queen's Plate.
7. Before the water level must be raised, the skiff will float.
8. The director treated Tina respectfully although she refused to cooperate.
9. The car headed straight toward them as they were too frightened to move.
10. Because we cannot be seated, the show has started.

practice 2

Rewrite the following sentences, removing *and* or *but* from between a main clause and a subordinate clause beginning with *which* or *who.*

1. The batter hit a foul ball and which was caught by a girl in the stands.
2. Pollution is a contemporary problem and which has no easy solutions.
3. *My Mother, the Doctor* was written by Joy Singer and who is also a physician.
4. The team showed spirit at first but which soon vanished as the score mounted against them.
5. Coleman makes a comfortable, roomy tent and which you will enjoy camping in with your family.
6. She was a singer of great talent but who never came up with a hit record.
7. Running is strenuous exercise but which many people find boring.
8. "The Love Song of J. Alfred Prufrock" is a dramatic monologue and which was written by T. S. Eliot.
9. Amelia Earhart was a renowned aviator but who vanished on a flight over the Pacific in 1937.
10. There's a new diving instructor at the Y and who took part in international competition two years in a row.

practice 3

Rewrite the following sentences. Correct the over-subordination in each sentence by leaving out unnecessary details, by changing clauses into phrases or single words, or by breaking the sentence into two sentences. Read your sentences aloud to make sure they sound natural.

1. After he left university he taught school in Winnipeg, where he met Shana West, who became his wife.
2. Georgia's car is a Chevrolet, which is made by General Motors, who also make Buicks and Cadillacs, that are more luxurious.
3. The toy, which had been made in Taiwan, where labor is low priced, was amazingly durable.
4. Over there is the lady who feeds the pigeons nuts which she buys from the vending machine that stands in the lobby.
5. In Morrisburg, Ontario, we visited Upper Canada Village which is a small village which has more than forty pioneer buildings some of which were moved there and some of which are reconstructions and which date back to the period from 1784 to 1867.

6. Diana has the part of Titania, who is the fairy queen in *A Midsummer Night's Dream,* which is a fantasy that was written by Shakespeare.
7. Here is a book which was written by Helen MacInnes, who has written all those fabulous cloak-and-dagger stories that are about spies and ladies who are in distress.
8. That's the trail bike which I bought for seventy-five dollars from the lady who owns the Shell gas station that is across the street from the school where I attended fifth and sixth grades.
9. I was living in a narrow row house which was behind the freight yards when I was six years old when I first began walking the three miles that lay between the house and the Monarchs' old ball park that squatted on the corner of Main and Third, where a high rise now stands.
10. While they were at Camp Centennial, the scouts decided to go on an overnight hike and, after they got permission from Pat Brown, who was their scouter and who was a lawyer back home, they started out for Kenesaw Mountain, which was a rugged peak that was in a valley which lay about five kilometres from camp.

application

Complete each of the following sentences by supplying the missing subordinate clause or the missing main clause. Choose the appropriate subordinating word from the following lists to begin the subordinate clauses.

Time: after, as, before, when, while

Condition: although, because, if, since, unless

Identification: who, which, that

Write out the complete sentence, using appropriate capitalization and punctuation.

1. The winner was Jean-Paul, _____. *(subordinate clause–identification)*
2. Since single-car accidents are mostly caused by speeding, _____. *(main clause)*
3. *(subordinate clause–condition)* _____ ticket prices in pro football are being raised again.

4. Although we ordered the material eight months ago, ____. *(main clause)*
5. *(main clause)* ____ unless the fog lifts.
6. *(subordinate clause–time)* ____ Carol enlisted in the army.
7. Oregano, basil, thyme and chives are the four spices ____. *(subordinate clause–identification)*
8. *(subordinate clause–time)* ____ the roof fell in.
9. While the acrobats tumbled and flipped through the air, ____. *(main clause)*
10. We will agree to cooperate ____. *(subordinate clause–condition)*

6 Double Subjects

Every sentence you write has a subject, as in the following sentences:

> **Elvis** threw his scarf to the audience.
>
> **The bottle** lay smashed alongside the curb.
>
> **She** struggled against the current.

Sometimes a sentence has a compound subject, that is, two subjects joined by *and.* You can write a sentence like this one, for example:

> **The Ford** and **the Pontiac** were favorites to win the race.

Sometimes a sentence doesn't seem to have any subject at all. You can write a sentence like this one:

> Turn the handle clockwise for three complete revolutions.

Even a sentence like this has a subject. Although the subject is not stated directly, the reader easily understands that the subject is the pronoun *you:* "You turn the handle . . ."

Every complete sentence, then, does have a subject. However, you should avoid writing a sentence with a double subject. A sentence like the following one is incorrect:

> *__My parents they__ don't quarrel much.

*The asterisk (*) identifies incorrect sentences.

This sentence has a subject—*my parents.* But that subject is repeated by the pronoun *they.* The result is a double subject. A double subject occurs whenever you insert the pronoun *he, she, it, we,* or *they* in a sentence immediately after the subject. This is an unnecessary repetition. To correct a double subject, simply drop the pronoun.

> **My parents** don't quarrel much.

Of course, a pronoun can be used as a subject if it has a clear referent in a preceding sentence. In the sentences that follow, *they* is used correctly as the subject of a sentence:

> **My parents** get along pretty well. **They** don't quarrel much.

Remember, don't immediately follow a subject with the pronoun *he, she, it, we,* or *they.* If you do, you've written a double subject.

practice 1

Write *double subject* after the number of each sentence containing a double subject. Write *correct* if the sentence is correct.

1. Paul he worked on his father's sod farm this summer.
2. Calvin and she wrote the school play.
3. My sisters they both went to McGill University.
4. The faucet it needs a new gasket.
5. Rona went waterskiing, and afterward she swam.
6. The rain it came down steadily, washing out the ball game.
7. The James boys and I went to the fair together.
8. Refik Shah he wanted a vest to wear with his suit.
9. As the treasurer, she had to balance the books.
10. Lorna she signed up for mechanical drawing.
11. The conductor wearing the green jacket he helped us with our luggage.
12. The three men entered the cafeteria.
13. Stacey and I we couldn't stay to see the end of the show.
14. That dishwasher it is always breaking down.
15. The three of us went to the soccer game at Empire Stadium.

practice 2

The following sentences all contain double subjects. Rewrite each sentence correctly.

1. The song it went on and on.
2. Three of the girls they forgot to bring water containers.
3. The police officer she asked to see my driver's licence.
4. The passengers they accepted the delay with good humor.
5. While asking Kim to dance, Porky he blushed.
6. Several of the players they protested the umpire's decision.
7. Because the humidity was low, the heat it did not bother me.
8. Floyd Cramer and Boots Randolph they appeared at the Queen Elizabeth Theatre.
9. That police officer with the long hair he fixed our car.
10. My brother he stood up for us.
11. Quickly the sled it shot down the icy hill.
12. The nurses they were cheerful.
13. All of us we couldn't wait until the ice cream was ready.
14. The gym it was decorated for the prom.
15. The lady with the gavel she is the auctioneer.

← application →

Write the following sentences, completing each one by supplying a subject.

1. __ wore purple slacks with a yellow sweater.
2. __ ordered a large pizza with olives and mushrooms.
3. __ collapsed with a loud roar.
4. Before noon __ was dry.
5. Swinging through the trees, __ overtook the safari.
6. __ cried softly in the moonlight.
7. __ and __ prevented a catastrophe.
8. __ typed my term paper, staying up till ten o'clock to finish.
9. __ was my big goal, my dream.
10. Because of the response, __ played another song.

Misplaced Modifiers

A modifier is a word or group of words that describes, qualifies, or otherwise identifies another word or group of words. A modifier may be a single word. It may be a phrase. Or it may be a clause. The words in color in the following sentences are modifiers:

The foundation was laid **quickly.** [single word]

Arguments **about rules** delayed the match. [phrase]

This story is about a woman **who flies balloons.** [clause]

When you use modifiers in a sentence—especially phrase and clause modifiers—try to place them as close as possible to the words they modify. For example, notice where the phrase and clause modifiers are placed in the following sentences:

The manikin **in the window** wears a dashiki. [phrase modifying a noun]

Please drive **with care.** [phrase modifying a verb]

The girl **throwing the javelin** is Elena Gorchakova. [phrase modifying a noun]

The lot **that is on the corner** could be a garden. [clause modifying a noun]

Using modifiers makes your writing clearer and more precise. But if you're not careful where you place them, you can easily confuse your reader. For example, notice the confusion caused by the misplaced modifier in this sentence:

*The guidebook says that one can visit the Peace Tower **on page thirty-seven.**

The phrase *on page thirty-seven* is a modifier. It is intended to modify *says.* But because of its position, it appears to modify *Peace Tower,* or possibly even *visit.* Although you can easily figure out what this sentence is supposed

*The asterisk (*) identifies incorrect sentences.

to mean, you shouldn't have to. The meaning is perfectly clear when the modifier is placed where it really belongs.

> The guidebook says **on page thirty-seven** that one can visit the Peace Tower.
>
> **On page thirty-seven** the guidebook says that one can visit the Peace Tower.

In most cases you'll avoid problems by putting modifiers as close as possible to the words being modified. Here are some specific situations that illustrate this general rule:

1. Place modifiers like *almost, even, ever, hardly, just, merely, nearly,* and *only* immediately before the words they modify.

> We could not **even** budge the fallen limb.
>
> **Even** we could not budge the fallen limb.
>
> The chief **only** talked to her.
>
> The chief talked **only** to her.

2. Place clause modifiers beginning with *who, which,* or *that* immediately after the words they modify.

> *She wrapped the scarf around her throat **which she had purchased at the garage sale.**
>
> She wrapped the scarf **which she had purchased at the garage sale** around her throat.

There are some situations, however, that require special rules for placing modifiers correctly.

1. Avoid inserting a modifier between the word *to* and the verb that follows, as in *to leave.*

> *The fire marshal asked the demonstrators *to* **quickly but calmly** *leave.*
>
> The fire marshal asked the demonstrators to leave **quickly but calmly.**

2. Avoid inserting a modifier between a verb and its object, unless the object itself is being modified. In the following sentence, *threw* is the verb; *a fastball and a change-up* is its object.

*She *threw* **with the same motion** *a fastball and a change-up.*

She threw a fastball and a change-up **with the same motion.**

3. Avoid inserting a modifier between a preposition and its object, unless the object itself is being modified. In the following sentence, *into* is the preposition; *the principal's office* its object.

*The boy walked *into,* **frightened,** *the principal's office.*
Frightened, the boy walked into the principal's office.

4. Avoid separating coordinate modifiers. Coordinate modifiers are modifiers that take the same form and modify the same words. They should be joined by *and, but,* or *or,* not separated.

***With fair weather** the Greenpeace should have an uneventful voyage **with good luck.**

With fair weather *and* **good luck,** the Greenpeace should have an uneventful voyage.

5. You can put phrases and clauses that tell *when, where,* or *how* at beginnings of sentences even though this may separate them from the words they modify.

When the final lap started, Cheryl moved past the front-runners.

In their hurry, the commuters failed to notice the old man selling flowers.

practice 1

In the following sentences, if the modifiers in heavy type have been correctly placed, write *correct* after the number. If they have been misplaced, write *misplaced* after the number.

1. She polished her boots until **almost** they looked new.
2. Every part must pass a thorough test **that goes into our stereo components.**
3. The farmer was **only** willing to sell part of his land.
4. The TV picture was relayed by the satellite **that orbited the earth.**

5. Corinne read a book about Winston Churchill **who is writing a term paper on the cold war.**
6. The ambulance should arrive **within minutes.**
7. We have spent our vacation camping in, **for several years,** Banff National Park.
8. **Before the sale ends,** rush to your nearest furniture dealer.
9. **Quietly** the astronauts waited **nervously.**
10. Jefferson set, **who was a so-called rabbit,** a fast pace.
11. The truck transporting the giraffes made it through, **just barely,** the underpass.
12. Twenty dollars is **nearly** not enough for this original oil.
13. The mouse rang the bell **without hesitating** and received a piece of cheese.
14. Munson, the catcher, threw **accurately** the ball to second base.
15. The committee voted to **unanimously** give this year's service award to Ms. Ramiriz.

practice 2

The following sentences contain misplaced modifiers. Rewrite each sentence, placing the modifier in heavy type correctly.

1. Mollusks, or shellfish, are widespread **which yield pearls.**
2. You should begin to, **if you want a summer job,** look now.
3. **Silently** the sub slipped beneath the waves **swiftly.**
4. The independent candidate **only** won one poll.
5. Ceremonies started just as we arrived **with "O Canada."**
6. The poems express a great love of liberty **by Burns.**
7. Some newscasters **hardly** show any interest in making their news programs entertaining.
8. The car spun out in the number three turn **that had been leading the race.**
9. Roberta Flack performed **brilliantly** her latest hit.
10. The house was built in, **although you would scarcely believe it,** the 1800s.

11. Several debates had occurred during the convention **concerning procedure.**
12. Carla can jump over, **though she's 150 centimetres,** a 160 centimetre high fence.
13. The audience seemed to enjoy the Mendelssohn violin concerto **sitting on the grass around the band shell.**
14. We were asked to **generously and unselfishly** contribute.
15. **Loudly** the calf bawled for its mother **frantically.**
16. One must consider, **when buying a ticket,** the price.
17. The docking of the spacecraft was a success **which required delicate manoeuvring.**
18. The runners **hardly** had enough stamina to complete the last mile at such a fast pace.
19. For, **completely exhausted,** sixteen hours the survivors slept.
20. I put the chair in the bedroom **that I had bought at the flea market.**

← application →

In the following paragraphs, there are ten sentences which contain misplaced modifiers. Rewrite each of the ten sentences, placing the modifiers correctly.

> Little League baseball is a popular program, but it's not a good one for the 8- to 12-year-olds. Here's something to as an alternative consider. Kids should play softball instead of baseball. Most kids are afraid of, slightly, a baseball. It's hard and it can hurt. A softball is bigger than a baseball. And one can't throw as fast a softball. Therefore, it's easier to catch and to solidly hit.
>
> What's the real fun of playing ball? The running, the throwing, and the hitting. The pitchers dominate Little League baseball who are usually the biggest and best athletes on the team. Everyone else stands mostly around. In a softball game, everyone is involved, from the first baseman to the right fielder.
>
> There's time to play, when kids are older, baseball. Then, with the experience of softball behind them, they'll be better prepared for the hardball game.

They'll have mastered that softball teaches the fundamentals. They'll also be better coordinated. Sports programs are good for those who only benefit. Adults have been force-feeding baseball in their enthusiasm to the subteens. It's time to put some fun back in the game.

8 Dangling and Squinting Modifiers

Modifiers are words, phrases, and clauses that describe, limit, or qualify other words in the same sentence. In the following sentences, the phrases and clauses in color are all modifiers:

The ad asked for a person **with a chauffeur's license.**

Use a cooking oil **that is high in polyunsaturates.**

The Minister of Finance spoke **about improving the country's economy.**

As I witnessed the ceremony, I could feel goose bumps growing.

When you use a modifier, be sure that it clearly refers to the specific word or words you want it to modify. Single-word modifiers, like adjectives and adverbs, don't usually cause trouble. But phrases and clauses used as modifiers are often misplaced, resulting in awkward or confusing sentences. Two common errors in placing modifiers are called *dangling modifiers* and *squinting modifiers.*

Dangling Modifiers A dangling modifier is a phrase or clause that cannot logically refer to any other element in the sentence. In the following sentence, the phrase in color is a dangling modifier:

***Unscrewing the radiator cap,** a blast of steam shot up.

Unscrewing the radiator cap is a modifying phrase. But there is nothing that it can logically modify in the rest of the sentence. Obviously it cannot modify *a blast of steam.* Thus, the phrase is left "dangling," a modifier with nothing to modify.

To correct a dangling modifier, you can rewrite the main part of the sentence, adding a word that can be modified by

*The asterisk (*) identifies incorrect sentences.

the dangling modifier and changing other words if necessary.

Unscrewing the radiator cap, **I** released a blast of steam.

In this sentence, *unscrewing the radiator cap* clearly modifies *I*. The verb was changed from *shot up* to *released* so that the meaning of the sentence remains about the same.

Another way to correct a dangling modifier is to rewrite the modifier, turning it into a complete clause.

As I unscrewed the radiator cap, a blast of steam shot up.

By adding the subject *I* to the phrase and by changing the form of the verb, you change the phrase to a complete clause.

How can you spot a "dangler"? Most dangling modifiers occur at the beginning of sentences, but sometimes they occur at the end. Also, they take several different forms. Often they are verbal phrases like these:

Running down the steep hill, . . . [participle]

To finish the project on time, . . . [infinitive]

Before **signaling** a left turn, . . . [gerund]

Sometimes dangling modifiers are elliptical clauses. That is, they are clauses in which the subject and the verb are omitted. Such clauses usually begin with words like *when, while, if, although*. These are elliptical clauses:

When only a small child . . . [The words *I was* are omitted.]

If bored and tired . . . [The words *you are* are omitted.]

Here are some examples of dangling modifiers that have been corrected. Notice where and how words have been added to the original sentences.

*By **crossing the Canso Causeway,** Nova Scotia's Cape Breton Island can be reached. [dangling gerund phrase]

By **crossing the Canso Causeway, *motorists*** can reach Nova Scotia's Cape Breton Island.

*The game was enjoyed **sitting in the box seats.** [dangling participial phrase]

The game was enjoyed by the ***people* sitting in the box seats.**

***To collect $200,** *Go* must be passed. [dangling infinitive phrase]

To collect $200, *you* must pass *Go.*

***While delivering papers,** a dog bit my ankle. [dangling elliptical clause]

While *I was* delivering papers, a dog bit my ankle.

Squinting Modifiers The word *squint* means "cross-eyed" or "looking in two directions at once." A squinting modifier is any modifier placed so that it could logically refer either to what comes before it or to what comes after it. The word in color in the following sentence is a squinting modifier; it "looks" in two directions at once:

*The teakettle that had been whistling **noisily** boiled over.

Does *noisily* describe the way the teakettle whistled or the way the teakettle boiled over? Either would make sense. There is nothing in the sentence to indicate the writer's meaning.

Squinting modifiers are relatively easy to correct. You can simply switch the modifier around, moving it to a different position in the sentence.

The teakettle that had been **noisily** whistling boiled over.

The teakettle that had been whistling boiled over **noisily.**

Or, you can rephrase the sentence.

The whistling teakettle **noisily** boiled over.

The teakettle whistled **noisily** before it boiled over.

practice 1

In the following sentences, identify the modifiers printed in heavy type. If the modifier is used incorrectly, write either *dangling modifier* or *squinting modifier* after the number. If the modifier is used correctly, write *correct* after the number.

1. **Typing rapidly,** the essay was finished by ten o'clock.
2. The judge agreed **in her chambers** to hear the witness.

3. **Driving on bald tires,** the highway was slippery.
4. You should read widely in order **to write well.**
5. Their parents tried **thoroughly** to understand.
6. The walls shake **when slamming the door.**
7. Several students **we know** failed the test.
8. **If grass is dry in the early morning,** watch for rain by nightfall.
9. Beth decided **independently** to conduct the experiment.
10. **Jogging through the forest preserve,** unseen creatures skittered through the grass.
11. The workers demanded **on the next day** to be paid.
12. **When sanded and varnished,** the boat will look like new.
13. **After completing high school,** my father will give me a job.
14. The race continued **although weakened by fatigue.**
15. They agreed **when both sides ceased bargaining** to call in a mediator.

practice 2

Rewrite each of the following sentences to correct the dangling modifier printed in heavy type.

1. The deadline was met **writing in a frenzy.**
2. **Peering through the binoculars,** the black-throated green warbler was sighted.
3. **Even when sliced,** I like mushrooms.
4. The waiter seated us **after waiting in line for thirty minutes.**
5. **To grow good tomatoes,** tall stakes are necessary.
6. University was a scary maze **when just a freshman.**
7. **For winning the game,** a trophy was presented.
8. **Being very expensive,** I did not buy a ticket to the Rolling Stones' concert.
9. Never race a car engine **before warming up.**
10. The race was won **breaking the school record of four minutes thirty-four seconds.**
11. **While swinging a scythe,** his back gave out.
12. **To make headcheese,** the brine should be strong enough to float a potato.
13. The alarm sounded **on opening the rear door.**

14. **Having raised the fares,** the cab company's income increased.
15. **Headed down the mountainside,** momentum built up.

practice 3

The following sentences contain squinting modifiers. These are printed in heavy type. Rewrite each sentence so that the squinting modifier clearly modifies only one element.

1. The board of health asked **after ten days** to be informed.
2. The superintendent promised **immediately** to reform the curriculum.
3. The tugboat puffed and groaned **laboriously** coming upriver.
4. The salesman spoke **feverishly** chomping on a rancid cigar.
5. Spectators who cry out **rudely** disturb the performers.
6. Anyone, boy or girl, who can field **well** merits a chance to play.
7. I was on a one-way street **by mistake** going the wrong way.
8. The reduced fares were for senior citizens **only** on weekdays.
9. Mrs. Vallely promised the staff **when the yearbook was finished** they'd have a party.
10. The red-faced man **on the corner** sold used cars.

← application →

Complete the following sentences, making up a basic sentence to go along with the modifier you're given. Make sure you don't end up with a dangling modifier.

1. Having grown up in the 1960s, ___.
2. ___ rolling through the median strip.
3. To receive permission for a field trip, ___.
4. If properly frozen, ___.
5. By mowing the grass infrequently and cutting it high, ___.
6. ___ when riding on a train.

7. In the elevator, ___.
8. ___ leaning against the rear wall of the garage.
9. To understand Alice Cooper, ___.
10. Before I graduate, ___.
11. Upon entering the twelfth grade, ___.
12. ___ when the clock began to strike.
13. Tied securely to the boat's mast, ___.
14. ___ although it was only nine-thirty.
15. When riding a motorcycle, ___.

9 Faulty Parallelism

When you write a sentence that lists things in a series, make certain that all the items listed in the series appear in the same form. That is, make them parallel. The following sentence contains faulty parallelism. In listing three favorite activities, the writer shifted needlessly from the gerunds *swimming* and *fishing* to an infinitive, *to lie.*

*I have always enjoyed **swimming, fishing,** and **to lie** in the sun.

The words in color all serve the same function. They are all items in a series of activities. Therefore, they should all be written in the same form. The writer could use all gerunds.

I have always enjoyed **swimming, fishing,** and **lying** in the sun.

Or the writer could use all infinitives.

I have always liked **to swim, to fish,** and **to lie** in the sun.

Which form you choose for expressing parallel ideas is not important. What's important is that you be consistent in using the same form for all the parallel ideas in a sentence.

Faulty parallelism can occur with many different grammatical forms. Here are some examples of faulty parallelism caused by needless shifts from one form to another:

*Grade 12 orientation week seems **worthwhile** and **a necessity.** [shift from an adjective to a noun]

Grade 12 orientation week seems **worthwhile** and **necessary.** [two adjectives]

*The asterisk (*) identifies incorrect sentences.

*Our camp counsellor told us **to clean** our tents, **to police** the area, and **that we should stack firewood.** [shift from infinitives to a clause]

Our camp counsellor told us **to clean** our tents, **to police** the area, and **to stack** firewood. [all infinitives]

*This fibreglass canoe **holds** four people, **requires** no paint, and **it paddles easily.** [shift from verbs to a clause]

This fibreglass canoe **holds** four people, **requires** no paint, and **paddles** easily. [all verbs]

*White-water kayaking requires **strength, courage,** and **the paddler must be experienced.** [shift from nouns to a clause]

White-water kayaking requires **strength, courage,** and **experience.** [all nouns]

Notice that faulty parallelism commonly occurs with elements joined by *and, but,* or *or* and with items in a series. Usually, among items in a series, it's the last item that is needlessly shifted.

You should also be alert for faulty parallelism when you use a pair of correlative conjunctions like *either . . . or, neither . . . nor, not only . . . but also,* and *both . . . and.* Be sure that the ideas following both conjunctions in the pair are expressed in the same form.

*Sheila *not only* **gave** of her time *but also* **she donated a good deal of money.** [shift from verb to clause]

Sheila *not only* **gave** of her time *but also* **donated** a good deal of money. [two verbs]

Neither* **speed *nor* **being elusive** was enough for Clyde to become a starting halfback. [shift from noun to participle]

Neither **speed** *nor* **elusiveness** was enough for Clyde to become a starting halfback. [two nouns]

*She wished *both* **to make** the honour society *and* **that she might earn a school crest.** [shift from an infinitive to a clause]

She wished *both* **to make** the honour society *and* **to earn** a school crest. [two infinitives]

practice 1

Read each of the following sentences carefully. If the elements in heavy type in the sentence are parallel, write *parallel* after the number. If the elements are not parallel, write *not parallel.*

1. Her redecorated bedroom was **colourful** and **a quiet place to study.**
2. Jackson promised **to be more careful** and **that he would work more slowly.**
3. Both **talent** and **being dedicated** are necessary if you want to become a professional dancer.
4. The ball **sailed** over the bushes, **rolled** under the fence, and **bounced** into the street.
5. **Raking** the beach, **scrubbing** the boats, and **sweeping** off the dock took Kathy most of the morning.
6. No one enjoyed **the haying, the milking,** or **to plow** except Bill.
7. She was not only **helpful** but also **charming.**
8. Pasquale is not only talented **as a pianist** but also **in acting.**
9. **To climb** the rope ladder, **to swing** out at the last moment, and **to ring** the bell at the top of the mast takes daring and skill.
10. She served us a dessert **that looked** like melted ice cream and **tasting** like egg custard.

practice 2

Rewrite each of the following sentences, correcting the faulty parallelism.

1. If you want my opinion, I'd rather go skiing than to sightsee.
2. The new girl in gym class can serve, can set up plays, and she spikes like a pro.
3. Eloise has always been easygoing, soft-spoken, and has patience.
4. No one was told to turn off the lights or that the doors should be locked.
5. To sing with a rock group and making lots of money was her big dream.

6. Brent wanted both to pass the test and getting a *C* in the course.
7. Making a final decision and to stick to it was something Marcia just couldn't do.
8. Run straight at the linebacker, give him a head fake, and then cutting toward the sideline.
9. My grandfather told me that honesty, hard work, and being thrifty made him successful.
10. Laurie worked quickly, carefully, and was efficient.
11. Miguel said that he would be back soon and for us to start work without him.
12. This melon is ripe, juicy, and has a sweet taste.
13. In many countries, having enough to eat, having a comfortable home, and to be adequately clothed are considered luxuries, not necessities.
14. The teachers all knew about "skip day" and that grads cut classes.
15. Gardenia always liked taking a nap after lunch and to eat an early dinner.

practice 3

Rewrite each of the following sentences to eliminate faulty parallelism. Make certain that the ideas linked by each pair of correlative conjunctions are expressed in the same form. The correlative conjunctions are in heavy type.

1. Istvan would like **either** to enlist in the navy **or** learning computer programming.
2. The ranger warned us **not only** to dress warmly **but also** that we must keep dry.
3. Everyone complained **both** of the heat **and** how humid it was.
4. Kris wanted **either** to launch a new model rocket **or** repairing the ones flown yesterday.
5. The professor wanted **neither** sympathy **nor** to be pitied.
6. The mechanic **not only** changed the spark plugs **but also** she adjusted the carburetor.
7. Rumors were spread **both** about the roller rink being closed for good **and** that it was just being remodeled.
8. The instructor teaches one **not only** how to right a capsized boat and how to read weather signs **but also** signalling for help.

9. Nealy's aunt was **either** very tired **or** a very sick person.
10. **Neither** writing letters **nor** to read them seemed worth the effort.

← application →

Some part of each of the following sentences is missing. Write each sentence, supplying a part that is both appropriate in meaning and parallel in form to the rest of the sentence.

1. He told us that Mexico is a beautiful country, that its people are very friendly, and ___.
2. To get a good education, ___, and to live to a ripe old age all seem to be important parts of the North American dream.
3. Most people agree that ___, smoking tobacco, and drinking alcohol are dangerous to one's health.
4. Standing in a long line to see a movie and ___ are two things that I absolutely refuse to do.
5. Soccer, they say, can teach anyone quickness, courage, and ___.
6. She could use the prize money either to travel around the world or ___.
7. I like to sit outside at night ___, watching the cars go by, and listening to all the night sounds.
8. Even my best friend won't admit that I'm lovely to look at, delightful to know, and ___.
9. Persons that spread rumors, persons that borrow but never lend, and ___ are at the bottom of my popularity list.
10. Studying arithmetic will teach you not only how to add the figures on a shopping list but also ___.
11. The inspector looked for polished shoes, for ___, and for pressed uniforms.
12. A toast: May your tires never go flat, may your clocks never stop, and ___.
13. We understood that to be young was good and that ___ was best of all.
14. I believe in good music, vigorous exercise, and ___.
15. Time is always ___, just, and impartial.

10 Shifts in Tense and Voice

When you write a sentence with two main verbs (a compound verb), be careful to use the same grammatical form for both verbs. Don't shift from one form to another in the same sentence. Needless shifts in verb form result in faulty sentences like this one:

> *The Prime Minister's bodyguard **pushed** their way through the crowd lining the street and **make** an open corridor to the hotel entrance. [shift from past tense to present tense]

Notice that this sentence uses two verbs to describe two parallel actions of the Prime Minister's bodyguard. The first verb, *pushed,* is in the past tense. But the second verb, *make,* shifts to the present tense. Since both actions described by these verbs take place during the same time, the form of the verbs should be the same.

> The Prime Minister's bodyguard **pushed** their way through the crowd lining the street and **made** an open corridor to the hotel entrance. [both verbs in past tense]

Once you establish the tense with the first verb in a sentence, any other verbs you use to describe parallel actions should take the same tense form as the first. Shifting back and forth from one tense to another can only confuse your reader.

Needless shifts in verb form may also involve shifts in voice. English has two "voices"—the active and the passive.

> A reporter **saw** the accident. [active voice]
>
> The accident **was seen** by a reporter. [passive voice]

Notice that while the two sentences mean exactly the same thing, their verbs take different forms. Notice, too, that the subject of the verb, *a reporter,* and the object of the verb, *the accident,* switch positions in the passive sentence. And the passive verb includes a form of *BE* as one of its parts. When you write a sentence containing two or more verbs, be consistent. Don't shift from active to passive within the same sentence.

Knowing the differences between active and passive verb forms will help you recognize a sentence that shifts from

*The asterisk (*) identifies incorrect sentences.

one to the other. Here is a sentence illustrating this kind of needless shift:

> *Because Cindy **had missed** breakfast, the early lunch **was appreciated.** [shift from active voice to passive voice]

Despite what this sentence seems to say, both verbs refer to Cindy. Cindy missed breakfast, and Cindy appreciated the early lunch. Because both verbs describe what Cindy did, they should both be in the same voice.

> Because Cindy **had missed** breakfast, she **appreciated** the early lunch. [both verbs in the active voice]

Notice that when the passive verb *was appreciated* is changed to the active verb *appreciated,* a subject for *appreciated* is added—the pronoun *she.* This new subject replaces the phrase *the early lunch,* which becomes the object of the active verb *appreciated.*

Here is another sentence rewritten to eliminate a needless shift in voice. Again notice that a subject is added to the second part of the sentence, one which refers back to the subject of the first part.

> *When the hikers **rounded** a curve in the mountain path, a peaceful valley **was seen** far below.
>
> When the hikers **rounded** a curve in the mountain path, **they saw** a peaceful valley far below.

In the rewritten sentence, *they,* referring back to *hikers,* has been added as the subject of the active verb *saw.* And the phrase *a peaceful valley* moves to a position after the verb, becoming its object.

Sometimes a shift in voice occurs in a compound sentence—two sentences joined by *and* or *but.*

> *The hikers rounded a curve in the mountain path**, and** a peaceful valley **was seen** far below.
>
> The hikers rounded a curve in the mountain path **and saw** a peaceful valley far below.

You don't always need to add a subject when you change the voice of the second verb in a compound sentence. *Hikers* is clearly the subject of both active verbs *rounded* and *saw.* So, there is no need to add the pronoun *they.*

Needless shifts in tense and voice are careless errors that happen when you're in a hurry. They can easily be avoided if you make a real effort to be consistent in your use of verb forms.

practice 1

If a sentence contains a needless shift in tense, write *tense* after the number of the sentence. If a sentence contains a needless shift in voice, write *voice.* If a sentence contains no needless shifts, write *correct.*

1. Private-eye stories make detective work seem glamorous, but actually such work was usually routine and boring.
2. Although Jesús Alvarez tried to get to the stadium on time, the kickoff was missed.
3. The mysterious package was opened, but nothing was found inside.
4. The coach always took the team to the same restaurant and buys them the same pregame meal.
5. Janice had to show us her new litter of puppies as soon as we pulled into the driveway.
6. Tracy shook the bottle vigorously, and sticky root beer was squirted all over us.
7. The horse rears and threw the jockey into the starting gate.
8. If you want an early refund, file your tax form as soon as you possibly can.
9. Sandra got upset, and the free throw was missed.
10. The bobsled rattled through the first curve and then, picking up speed, shoots down the short straightaway.

practice 2

Rewrite each of the following sentences to eliminate a needless shift in tense.

1. Mama Lena prepared the sauce for the pasta and puts the bread into the oven to bake.
2. Emily always wants to stay up for the late movie but always fell asleep watching it.
3. Before Norbert had a chance to take his guitar out of the case, Bridget's mom says she didn't care for guitar music.

4. If you buy a good pickup cartridge, you got better stereo sound even if your set isn't new.
5. She put some money aside for emergencies but still comes up short at the end of the month.
6. We can make it to the beach if seven people rode in your Volks.
7. Glenda and Bev planted a big garden and then give all the vegetables to their friends.
8. Hiroko didn't like the fish but eats it to be polite.
9. She comes into the room, sits down, and did not say a word.
10. If I hadn't seen it with my own two eyes, I will not believe it.
11. The tornado roared above the town but mercifully doesn't touch down.
12. For years, his breakfast never varied; it consists of two sausages and a bowl of fruit.
13. Yesterday Janet told us that she has passed the course.
14. The committee approved the resolution, as both Liberals and Conservatives vote in favor of it.
15. Helen Jackson was an early advocate of Indians' rights in the United States and becomes known as the Harriet Beecher Stowe of the Indians with the publication of her novel *Ramona* in 1884.

practice 3

Rewrite each of the following sentences to eliminate a needless shift in voice.

1. Because Ray stopped to buy a morning paper, his train was missed.
2. Jo jumped when the door slammed, and the tray of dishes was dropped.
3. We painted the outside of the canoe dark blue, but the inside was painted eggshell white.
4. Because her car had more gas in the tank, it was driven instead of mine.
5. Even though Aguillar trailed the field for the first three laps, the race was won by fifty metres.
6. She got interested in sewing at school, and a course in tailoring was taken during the summer.

7. When Angie poked into a trash heap behind an abandoned farmhouse, a valuable antique bottle was found.
8. The audience rushed the stage, and the line of ushers blocking the aisles was broken through.
9. When we visited the auto plant, a good deal about assembly-line techniques was learned.
10. Ludmilla kept the best pieces in her collection, but the less valuable ones were sold.
11. When you finish high school, your next step in life must be decided.
12. The bicyclists reached the river, and then the bridge was crossed.
13. They abandoned ship, and not until after ten days adrift were they rescued.
14. On Wednesday the social studies class visited city hall, and on Thursday the district court was seen in action.
15. As you listen to this album, all your favorite rock stars will be heard—Elton John, David Bowie, Eric Clapton, and many, many others!

← application →

Rewrite the following paragraph, revising it to eliminate needless shifts in the tense and voice of the verbs.

> Harry Houdini was a famous magician who performs surprising feats of escape. To perform one popular trick, Houdini produced a twenty-foot length of rope, and a handkerchief was borrowed by him from the audience. Then Houdini asked for a volunteer from the audience who came up on the stage, and the handkerchief was tied around Houdini's wrists by this volunteer. The volunteer next looped the rope through Houdini's arms, and both ends were held onto. Houdini made a quick move The rope passed through the loops and falls to the floor. Many people saw the trick, but its mystery could not be solved. Finally, after many years, Houdini revealed the secret to his friends, and it was taught to them.

Using the Right Words

11 Subject and Object Pronouns

Use the personal pronouns *I, you, he, she, it, we,* and *they* as subjects for sentences. The pronouns in these sentences appear in typical subject positions:

> **I** am ready, and **he** is, too.
>
> **She** finished, but **they** didn't.
>
> **We** hope that **you** will agree.

Use the personal pronouns *me, you, him, her, it, us,* and *them* as objects in sentences—that is, as direct objects, indirect objects, or objects of prepositions. The pronouns in these sentences appear in typical object positions:

> Martha called **him.** [direct object]
>
> Tom gave **her** a present. [indirect object]
>
> This news is just between **you** and **me.** [object of preposition]

Use the pronouns *I, you, he, she, it, we,* and *they* after forms of *BE* like *is, am, are, was, were, be,* and *been.*

> It **is I** who wrote the letter.
>
> It **was she** who answered the telephone last night.
>
> It **must have been they** who left the message.

In speaking, expressions like *It's me* and *It's him* are appropriate and are frequently used, but in most writing you should use the more formal *It is I* or *It is he.*

Choosing the right pronoun is sometimes a problem when the pronoun appears in phrases like *Marie and I, Steve or her.*

> **Marie and (I, me)** saw the accident.
>
> This present is for **Steve or (she, her).**

In the first sentence, should you use *I* or *me?* To decide which pronoun to use, try each part of the phrase separately, testing both pronouns like this: *Marie saw the accident, I saw the accident, Me saw the accident.* You wouldn't say **Me saw,* so *Marie and I saw the accident* is the right choice.

In the second sentence, should you use *her* or *she?* Try each part of the phrase separately, testing both pronouns

*The asterisk (*) identifies incorrect sentences.

like this: *This present is for Steve, This present is for she, This present is for her.* Since you wouldn't say **for she,* the right choice is *This present is for Steve or her.*

practice 1

Rewrite each of the following sentences, choosing the correct pronoun from the words in parentheses.

1. Elaine gave (we, us) the report.
2. Nobody understands (I, me).
3. Joe called (they, them) about the game.
4. Ms. Greenburg talked with (he, him) on the bus.
5. (They, Them) are the ones Beth wants.
6. Is that package for (I, me)?
7. Jeff finally told (she, her) the whole story.
8. Otto thought (we, us) would arrive by ten.
9. John handed (he, him) the racket.
10. Do you think (she, her) will win the prize?

practice 2

Rewrite each of the following sentences, choosing the correct pronoun for formal written style.

1. It must have been (she, her) who did that.
2. It is (they, them) who must take the responsibility.
3. Could it have been (he, him) at the door?
4. Do you think it was (she, her)?
5. It is (I, me).
6. It wasn't (he, him).
7. I think the winner will be (she, her).
8. Are you sure the players were (they, them)?
9. No one can believe it was (we, us) who won.
10. Do you think it was (I, me) who said that?

practice 3

Rewrite each of the following sentences, choosing the correct pronoun to complete the phrase.

1. The hike leaders will be Dave and (I, me).
2. Just between you and (I, me), I don't believe it.
3. Paul and (I, me) are going to take the bus.
4. Janie usually eats lunch with (she, her) and Barb.
5. Either Bob or (he, him) will be the next captain.
6. It wasn't Jill or (I, me) who woke the children.
7. David made cookies for Betsy and (he, him).
8. Kevin and Dick have challenged you and (I, me) to a game.
9. Leonid and (she, her) are leaving now.
10. Did you and (he, him) finish early?
11. Before you and (she, her) leave, call home.
12. Are you going with Judy or (I, me)?
13. Bill and (I, me) have something to tell you.
14. Could it have been Joe and (he, him) who did that?
15. (They, Them) and Al will buy the food.
16. We sat next to Sally and (she, her) at the game.
17. These books are for you and (they, them).
18. The message is from Tom and (he, him).
19. Pete and (they, them) will help sell tickets.
20. We left Olga and (they, them) at the bus stop.

application

Write sentences of your own, using the following phrases correctly.

1. Dave and I
2. Dave and me
3. you and her
4. you and she
5. Sue or he
6. Sue or him
7. Ralph and us
8. Ralph and we
9. Nan and them
10. Nan and they

12 Personal Pronouns After *Than* and *As*

Sentences with phrases like *than him* and *as he* often contain incomplete comparisons. That is, some unnecessary words of the comparison have been omitted. If you use a personal pronoun in such a comparison, choose the pronoun you would use if the comparison were completely stated.

> Karen is almost strong **as he.** [as **he** is strong]

In this sentence the pronoun *he* acts as a subject, even though the words *is strong* have been omitted. Say the sentence with these words in it. You can hear that *he* is the right choice because you wouldn't say **him is strong.*

> This jacket fits you better **than him.** [than it fits **him**]

In this sentence the pronoun *him* is acting as direct object, even though the words *it fits* have been omitted. Since you wouldn't say **it fits he,* you know *him* is the right choice.

Sentences like *Karen is almost as strong as him* are frequently used in speaking, but in most writing you should use the pronoun that would be needed if the comparison were complete.

Sometimes either a subject pronoun or an object pronoun seems to fit. Your choice of pronoun will determine the meaning of the sentence.

> Anita helped Erik as much **as I.** [as **I** helped Erik]
> Anita helped Erik as much **as me.** [as she helped **me**]

practice 1

The following sentences have been written with the comparison fully stated. Rewrite each sentence, adding the correct pronoun from the words in parentheses.

1. Joanne reads faster than (she, her) does.
2. Mike likes chili as much as (I, me) do.
3. Mike likes Betsy as much as he likes (I, me).
4. Pat's sister plays chess as well as (she, her) does.
5. Nobody works harder than (we, us) do.
6. Carla takes longer to eat lunch than (I, me) do.
7. Dad enjoys cooking as much as (we, us) do.
8. I liked that book more than (he, him) did.
9. Pete looks more like you than he looks like (I, me).
10. Pete looks more like you than (I, me) do.

*The asterisk (*) identifies incorrect sentences.

practice 2

Rewrite each of the following sentences, supplying the word or words that make the comparison complete.

1. Ted likes Janie as much as we.
2. Ted likes Janie as much as us.
3. Mary admires Ms. Ohashi more than they.
4. Mary admires Ms. Ohashi more than them.
5. Joe listens to his brother more than she.
6. Joe listens to his brother more than her.
7. Mr. Berman paid you more than I.
8. Mr. Berman paid you more than me.
9. Jane praised Judy as much as he.
10. Jane praised Judy as much as him.

practice 3

Read each sentence, completing the comparison in your mind. Then rewrite each sentence, adding the correct pronoun from the words in parentheses.

1. Sometimes you're as lazy as (I, me).
2. Carol likes fudge as much as (he, him).
3. This color suits you better than (I, me).
4. Mr. Singh has a larger garden than (we, us).
5. Nobody understands our cat better than (I, me).
6. Doug felt worse than (they, them).
7. You were sick longer than (she, her).
8. Algebra is easier for Kathy than (he, him).
9. Speaking in public is harder for Bill than (she, her).
10. Julie practices the guitar as much as (he, him).

← application →

Write sentences of your own, using the following phrases and choosing the pronoun from the words in parentheses.

1. as fast as (he, him)
2. stronger than (I, me)

3. takes me longer than (they, them)
4. almost as tall as (she, her)
5. as talented as (he, him)
6. easier for you than (I, me)
7. as late as (we, us)
8. fits you better than (she, her)
9. as enthusiastic as (they, them)
10. much slower than (we, us)

13 Who and Whom

Use the pronoun *who* for the subject in questions like these:

> **Who** has finished? [subject of *has finished*]
>
> **Who** did it? [subject of *did*]
>
> **Who** was calling? [subject of *was calling*]

Use the pronoun *whom* for a direct object or for an object of a preposition. Notice that in questions the object pronoun *whom* appears at the beginning.

> **Whom** did you see? [direct object of *see*]
>
> To **whom** did you send it?
> **Whom** did you send it to? } [object of preposition *to*]

Use *who* in questions with forms of *BE* like *is, am, are, was, were, be,* and *been.*

> **Who is** she?
> **Who was** that?
> **Who were** the actors?
> **Who could be** the leader? } [with *BE*]

Often, you must choose between *who* and *whom* to begin a clause within a sentence. Your choice depends on how the pronoun is used in its own clause, not on the structure of the entire sentence.

If a subject for the clause is needed, use *who.*

> Can you tell us **who** has finished? [subject of *has finished* in the clause *who has finished*]

If an object in the clause is needed, use *whom.*

Do you remember **whom** you saw? [direct object of *saw* in the clause *whom you saw*]

We called the man to **whom** you sent it.

We called the man **whom** you sent it to. [object of the preposition *to* in the clause *to whom you sent it*]

If the verb in the clause is a form of *BE,* use *who.*

I know **who** she is. [with *BE* in the clause *who she is*]

The difference between *whoever* and *whomever* is the same as the difference between *who* and *whom.* That is, use *whoever* for a subject and with forms of *BE* like *is, am, are, was, were, be,* and *been.*

Give this to **whoever** opens the door. [subject of *opens* in the clause *whoever opens the door*]

Give this to **whoever** is the winner. [with *BE* in the clause *whoever is the winner*]

Use *whomever* for a direct object.

Give this to **whomever** you choose. [object of *choose* in the clause *whomever you choose*]

Sometimes a phrase like *do you think* or *I believe* interrupts a sentence or clause containing *who* or *whom.* Such interrupting phrases don't affect the choice of pronoun. You may find it helpful to think of the sentence or clause without the interrupting phrase in order to choose between *who* and *whom.*

This is the person **who** ~~I believe~~ **can do the job.** [subject of *can do* in the clause *who can do the job*]

This is the person **whom** ~~Jan said~~ **you mentioned.** [object of *mentioned* in the clause *whom you mentioned*]

In speaking, *who* rather than *whom* is often used at the beginning of a sentence or clause, such as *Who did you give it to?* or *I know who you mean.* But in most formal writing, you should maintain the distinction between *who* as subject and *whom* as object.

practice 1

Rewrite each of the following questions, choosing *who* or *whom* to complete each question in the formal style.

1. ___ is that woman in the white lab coat?
2. ___ did you help?
3. ___ is going to bring the food?
4. ___ have you told about the contest?
5. ___ knows how to do this assignment?
6. ___ was the man in the dashiki?
7. To ___ did you give the message?
8. ___ did he ask for directions?
9. ___ should I address this application to?
10. ___ are they going to hear at the conference?

practice 2

Each of the following sentences contains a clause in heavy type with the word *who* or *whom* omitted. Rewrite each sentence in the formal style, adding *who* or *whom* to each clause.

1. Matt met the woman ___ **will speak on air pollution.**
2. Can you tell me ___ **he is?**
3. You can go with ___**ever you like.**
4. Did Judy tell you ___ **she photographed?**
5. ___**ever did this** will be sorry.
6. ___**ever they elect** will be a very busy person.
7. Please tell me ___ **you have chosen.**
8. Do you know **with ___ they are traveling?**
9. This notebook should be returned to ___**ever owns it.**
10. "People ___ **live in glass houses** shouldn't throw stones."

practice 3

Rewrite each of the following sentences, adding the pronoun *who* or *whom* to complete each sentence in the formal style. Then draw a line under the pronoun and its clause.

1. I can't imagine ___ it could be.
2. Sharon will give the message to ___ever answers the telephone.
3. The students ___ Phil met yesterday are all in grade twelve.
4. Have you heard ___ will be at the party?
5. I have no idea ___ they have invited.
6. Has anyone told Joe ___ sent the mysterious package?
7. I don't know to ___ I should report the accident.
8. ___ever Dan appoints will be responsible for the tickets.
9. Helen can't remember ___ told her that.
10. The woman ___ Mary spoke to is her science teacher.

practice 4

The following sentences contain interrupting phrases like *do you think*. Rewrite each sentence, adding *who* or *whom*. Enclose each interrupting phrase in parentheses.

1. ___ do you suppose was at the rally?
2. Tell me ___ in your opinion should be captain.
3. Leroy is the student ___ I believe you have already met.
4. Mom talked with the woman ___ she hopes will be our next mayor.
5. Give the award to ___ever in your opinion has earned it.
6. We saw the same hikers ___ I think you saw last week.
7. ___ does he think will finish first?
8. Tom is the one ___ she is certain will be elected.
9. ___ in Jack's opinion did they mean?
10. She is the one ___ according to the paper was an eyewitness.

← application →

Write sentences of your own, using the following phrases and choosing the pronoun from the words in parentheses.

1. the swimmer (who, whom) broke the record
2. the singer (who, whom) I like best
3. the teacher to (who, whom) you gave your report
4. (whoever, whomever) finishes first
5. (whoever, whomever) you choose
6. the one (who, whom) I hope will win

7. the athlete (who, whom) I think Ann met
8. the woman (who, whom) was the treasurer
9. the actor (who, whom) Paul saw
10. the adviser with (who, whom) you talked

14 Who, Which, and That

In general, use the pronouns *who* and *whom* to refer only to people.

> The woman **who** is giving a speech is a Member of Parliament.

> Will the swimmers **whom** you saw be competing in the meet?

Use the pronoun *which* to refer only to animals or things.

> Cowboys need horses **which** can start, stop, and turn quickly.

> That factory, **which** is empty, is being torn down.

However, when an animal is referred to by name, *who* and *whom* are sometimes used: *My dog Dina,* **who** *is almost ten years old, is getting lame.*

Use the pronoun *that* to refer to either animals, things, or people.

> The cat **that** is sleeping on the porch belongs next door.

> Has anyone seen the book **that** I was reading?

> Have you met the basketball player **that** Tom knows?

practice 1

Rewrite each of the following sentences, choosing *who, whom, which,* or *that* from the words in parentheses.

1. Do you know the white-haired man (which, who) feeds the pigeons?
2. The bike (that, who) I was hoping to buy has been sold.

3. We all remember the dancers (which, who) performed in the park last summer.
4. Nobody understands the math assignment (that, who) is due tomorrow.
5. Wolves, (which, who) have strong family ties, often mate for life.
6 What did you think of the discussion (who, that) followed the meeting?
7 Will anyone (which, who) is interested please let me know?
8 Ice hockey, (which, who) developed in Canada, probably came from the older game of field hockey.
9. Did you see the people (who, which) gathered outside the mayor's office?
10. The Cabot Trail (who, which) extends for 310 kilometres on Cape Breton Island is one of the most scenic highways in Canada.
11. The passengers (whom, which) the conductor helped from the train were not injured.
12. Customers (who, which) are barefoot are not allowed in the store.
13. In 1976 people from around the world flocked to Montreal (which, who) was the site of the Summer Olympic Games.
14. The band members (whom, which) Carla telephoned will all attend the rehearsal.
15. Did you like the pizza (who, that) Joe and Julian made?

← application →

Write sentences of your own, using the following phrases and choosing the appropriate pronoun from the words in parentheses.

1. the survivors (whom, which) the reporter interviewed
2. students (who, which) like photography
3. buildings (who, that) are fire hazards
4. the program (whom, that) Nina likes best
5. dogs, (who, which) are normally friendly,
6. plants (who, that) need a lot of sunlight
7. school libraries, (which, who) are closed on weekends,
8. your friend (which, who) collects stamps
9. book reports (who, that) are neatly typed
10. my cousin, (which, whom) you met,

15 Pronoun Agreement

You use a pronoun to represent, or refer to, another word—its **antecedent.** A pronoun and its antecedent may be in the same sentence or in two separate sentences. In the following examples, *she* is the pronoun and *Sandy* is the antecedent:

> When **Sandy** gets home from school, **she** likes to play basketball.
>
> When **she** gets home from school, **Sandy** likes to play basketball.
>
> **Sandy** really enjoys basketball. **She** plays every afternoon.

When you write, make certain that your pronouns agree with, or match, their antecedents in three ways: number, person, and gender.

Number A pronoun should agree with its antecedent in number. That is, the pronoun should be singular if its antecedent is singular, and plural if its antecedent is plural.

> I like **pizza,** especially when **it** has anchovies. [singular]
>
> **Oranges** are delicious, especially when **they** are ripe. [plural]

In formal writing, certain indefinite pronouns like *somebody, everyone, anybody,* and *nobody* are considered singular. Use a singular pronoun to refer to them.

> **Somebody** left **his** bike unlocked.
>
> **Nothing** looks the way **it** used to.
>
> Does **everyone** understand what **she** is supposed to do?

In speaking, however, plural pronouns are often used to refer to these indefinite pronouns: *Everybody should pay their dues promptly.*

Watch for errors in number agreement with sentences like this:

*If a **student** registers late, **they** have to pay a fine.

Here the antecedent *student* is singular, even though it means "students in general." You can correct such an error by changing either the antecedent or the pronoun to make the two agree in number.

If a **student** registers late, **he** has to pay a fine. [both singular]

If **students** register late, **they** have to pay a fine. [both plural]

Use a plural pronoun to refer to two singular antecedents joined by *and.*

We spoke to **Beth and Joe** as soon as **they** arrived.

Use a singular pronoun to refer to two singular antecedents joined by *or* or *nor.*

Karen or Sara will give **her** report tomorrow.

Neither **Sam nor his brother** rode **his** bike today.

Person A pronoun should also agree with its antecedent in person. That is, both pronoun and antecedent should be either first person, second person, or third person.

First-person pronouns refer to the person speaking—*I, me, my, mine*—or to a group including the speaker—*we, us, our, ours.*

Hello, **this** is **Betsy** speaking. **I** hope you can hear **me.**

Second-person pronouns refer to persons spoken to.

You should be more careful of **your** bike, **Dave.**

Third-person pronouns refer to persons or things spoken about—*he, him, his, she, her, hers, it, its, they, them, their, theirs.*

When **Sally** saw the **children, she** called to **them.**

*The asterisk (*) identifies incorrect sentences.

Watch for errors in person agreement with sentences like this:

> *If a **student** registers late, **you** have to pay a fine.

Here the antecedent *student* is third person—someone you are talking about. So, you shouldn't use a second-person pronoun like *you.* You can change either the antecedent or the pronoun to make the two agree in person.

> If a **student** registers late, **he** has to pay a fine. [both third person]
>
> If **you** register late, **you** have to pay a fine. [both second person]

Gender The pronouns *he, him, his, she, her, hers, it,* and *its* should agree with their antecedents in one more way—in gender, or sex. Use *he, him,* and *his* for masculine antecedents; *she, her,* and *hers* for feminine antecedents; and *it* and *its* for neuter antecedents.

> **Diane** doesn't want this **book. She** has already read **it.**
>
> The new **jacket** fit **Max,** and **its** color suited **him** well.

practice 1

The following sentences are all written correctly. The antecedents of pronouns are printed in heavy type. Rewrite each sentence and draw an arrow from the pronoun to its antecedent.

1. If **Marie and Carol** are there, Dee can sit with them.
2. Finally Steve checked the **gas tank.** It was empty.
3. If it is frightened, a **dog** may bite.
4. **Someone** must have lost her wallet.
5. When **people** are in a hurry, they don't like delays.
6. The **boys** left early. They couldn't hear the speaker.
7. Neither **Roger nor Wayne** remembered to pick up his books.
8. **You** won't be able to vote if you miss the meeting.
9. Doesn't **anyone** want to sell his ticket to the game?
10. If **Marie or Carol** is there, Dee can sit with her.

practice 2

Rewrite the following sentences, choosing the pronoun or phrase in parentheses that agrees in number with its antecedent.

1. A cat often purrs when (it is, they are) contented.
2. Have you tried the chocolate cookies? Marty made (it, them).
3. Celia helped pick strawberries. (It was, They were) delicious.
4. All students should complete the assignment by Friday, or (they, he) will lose credit.
5. Unless drivers display parking permits on the rear window, (they, he) will be ticketed.
6. A student must use (their, his) ID card to be admitted.
7. If a person expects to play the guitar well, (they, he) must be willing to practice often.
8. When a child uses the wading pool, (he, they) must be accompanied by an adult.
9. If anyone wants to usher at graduation, (they, he) should see Ms. Reilly.
10. Jo Ann hunted everywhere for the badminton net. (They were, It was) in the attic.

practice 3

Each of the following sentences contains an indefinite pronoun as antecedent. Rewrite each sentence in formal style, choosing the pronoun in parentheses that agrees in number with its antecedent.

1. Did you tell everyone to wear (his, their) hiking boots?
2. If someone really hunts for that book, (they, he) will find it.
3. No one wants to give up (his, their) place by the front window.
4. Each of the women turned in (their, her) report on air pollution.
5. Anyone who wants to cook (his, their) own breakfast is free to do so.

6. Everybody in the girls' glee club should sign (their, her) name in the guest book.
7. If anyone has information about the accident, (they, she) should report it immediately.
8. Everyone on the girls' soccer team will buy (her, their) own bus ticket.
9. Somebody left (their, his) bicycle in the driveway.
10. After the checkroom tickets were lost, no one knew where (her, their) coat was.

practice 4

In the following sentences, antecedents are joined by *and, or,* or *nor.* Rewrite each sentence, choosing the pronoun in parentheses that agrees in number with its antecedent.

1. Have you seen the paste and masking tape? I can't find (them, it).
2. Neither Mr. Stevens nor Mr. Feldman brought (their, his) tennis racket.
3. Karen and Edie will show slides of (their, her) hiking trip.
4. Mom and Aunt Lou don't want to give up (her, their) consumer-education class on Saturday.
5. I think Joe or Tom will lend us (his, their) bike.
6. If you see Anne or Betsy, ask (them, her) to call me.
7. My cousin and his wife took two Airedale puppies with (him, them) on vacation.
8. Neither Sonja nor her sister remembered (her, their) house key.
9. Water and grease have left (its, their) mark on this old wallpaper.
10. Either my cat or yours has spilled (its, their) food on the kitchen floor.

practice 5

In seven of the following sentences, the pronoun in heavy type does not agree in person with its antecedent. Rewrite these sentences, replacing the pronoun with one that agrees with its antecedent. In three of the sentences, the pronoun in heavy type is correct. Write *correct* after the number of these sentences.

1. If somebody wants to sign up for the girls' softball team, **you** had better do it today.
2. When people go to the movies, **they** should try not to disturb others.
3. If a person wants to use the tennis courts, **you** must wear sneakers.
4. When a student takes a test, **you** should try to be relaxed.
5. After a club member uses the kitchen, **you** should clean everything up.
6. You should be careful in using these tools, or **you** might hurt yourself.
7. No one should work so hard that **your** health suffers.
8. Often a bike rider forgets that **you** should use hand signals in traffic.
9. If anyone is going to be late to Girls' Club, **you** should telephone Mrs. Metz.
10. If someone wants to leave early, **he** may do so.

application

Rewrite the following paragraphs, choosing from each set of words in parentheses the right word or words for formal style.

Not everyone is interested in auto mechanics or in doing minor maintenance and repairs on (his, their) own car. Such a person may feel that (his, their) car will perform better if a trained mechanic cares for (it, them). But even someone who has no interest in changing a flat tire or checking the oil and water still has an obligation to understand certain things about (his, their) car. (He, They) should know enough about the way (it operates, they operate) to be able to talk intelligently with the mechanic. The car owner should be able to describe what (he thinks, you think) is wrong and to explain what (he wants, you want) done.

Of course this situation applies to both men and women, since both may be car owners and both will want (her, their) car to perform at (its, their) best. Today many car owners are finding that a more thorough understanding of what is under the hood will help (you, them) avoid two things—being overcharged and having inferior work done on (your, their) car.

16 Pronoun Reference

When you use pronouns, keep in mind that they always seem to refer to the nearest noun, even though the nearest noun may not be the real antecedent. This tendency can cause two kinds of errors in pronoun reference—wrong reference and unclear reference. A third kind of pronoun error—no reference—occurs when there is no specific antecedent for the pronoun to refer to.

Wrong Reference In sentences containing wrong reference, a pronoun refers to the wrong antecedent, usually because the wrong antecedent comes between the pronoun and its real antecedent. The results are often unintentionally humorous.

*Ted took off his wet shoes, wiggled his bare **toes,** and put **them** in his knapsack. [wrong reference]

To correct this kind of pronoun error, rewrite the sentence, placing the pronoun as close as possible to its real antecedent.

Ted took off his wet **shoes,** put **them** in his knapsack, and wiggled his bare toes.

Sometimes, in rewriting, you will have to switch the positions of antecedent and pronoun and change the order of the sentence parts in which they appear. This change is usually necessary in sentences that begin with words like *if, when, although, since,* and *while.*

*The asterisk (*) identifies incorrect sentences.

*If the meat is too tough for the **kitten,** cut **it** into little pieces. [wrong reference]

Cut **the meat** into little pieces if **it** is too tough for the kitten.

Unclear Reference In sentences containing unclear reference, a pronoun seems to refer to any of several antecedents instead of just one.

*__Nan__ doesn't like to play tennis with her **sister** because **she** always wins. [unclear reference]

Who always wins? Does the pronoun *she* refer to *Nan* or to *sister?* To correct this kind of error, you may need to replace the pronoun with its specific antecedent.

Nan doesn't like to play tennis with her sister because **Nan** always wins.

Or, you may need to rewrite the sentence to move the pronoun closer to its real antecedent.

Because **Nan** always wins, **she** doesn't like to play tennis with her sister.

Sometimes, in order to make the antecedent of the pronoun clear, you may need to rewrite a sentence to include a direct quotation.

*__Jim__ told **Don** that **he** might miss band practice. [unclear reference]

Jim told **Don,** "**You** might miss band practice."

Jim said, "Don, **I** might miss band practice."

No Reference In sentences containing no reference, a pronoun has no specific antecedent at all but refers in a loose way to a general idea. Compare these two sentences:

No one finished the **assignment, which** was too long. [specific]

*No one finished the assignment, **which** was unfortunate. [no reference]

In the first sentence, the pronoun *which* has a specific antecedent, *assignment.* But in the second sentence, *which* seems to refer to the general idea expressed in the first part of the sentence, *no one finished the assignment.* Here are two possible revisions:

> **That no one finished the assignment** was unfortunate.
>
> **Unfortunately,** no one finished the assignment.

In each set of sentences that follows, the first example contains a pronoun with no specific antecedent. In the second example, the problem of no reference has been solved by eliminating the pronoun and replacing it with specific words.

> ***They** say more rain is on the way. [no reference]
>
> **The weather bureau** says more rain is on the way.

> Physical education programs are changing. ***This** will provide for more coed gym classes. [no reference]
>
> Physical education programs are changing. **These changes** will provide for more coed gym classes.

> Last week I had three tests and a book report due. ***It** was much better this week. [no reference]
>
> Last week I had three tests and a book report due. **My schedule** was much better this week.

Pronouns like *it, they, this, that,* and *which* are often used in this general way in speaking. In writing, however, you should be careful to include specific antecedents or to eliminate pronouns with no reference.

practice 1

The following sentences contain pronouns with wrong reference. Rewrite each sentence so that the pronoun in heavy type refers specifically to the antecedent in heavy type.

1. If this **medicine** doesn't help the dog, throw **it** away.
2. As we ate **chili,** we watched the sunset, enjoying **its** hot, spicy flavor.

3. The cabinet minister greeted the **reporters,** faced the TV lights, and hoped **they** wouldn't ask about the fuel shortage.
4. Last night the Debating Club **members** spoke on "White Collar Criminals." Only six of **them** were present.
5. As he poured the iced **tea,** Hank bumped into the refrigerator door and spilled **it.**
6. Daniel picked up his **notes,** turned to his classmates, and placed **them** on the speaker's desk.
7. When these **reports** are ready for the committee members, **they** should be stapled and bound.
8. Beth got into the **car,** adjusted the mirror, and turned **it** on.
9. Since last year's **coats** are too small for the twins, **they** should be given away.
10. If the bike **paths** are confusing to cyclists, **they** should be clearly marked with yellow paint.

practice 2

The following sentences contain pronouns with unclear reference. Rewrite each sentence so that the pronoun in heavy type refers specifically to the antecedent in heavy type. Or, if necessary, rewrite the sentence, replacing the pronoun with its specific antecedent.

1. Ben talked to **Raphael** about **his** trying out for the team.
2. The **captain** called the judge to tell him that it was **his** responsibility to enforce the laws, not to make them.
3. "Anything Grows" is the **name** of our school garden club, but I don't like **it.**
4. Carla studied **photography** before she went mountain climbing. She said **it** was hard.
5. **Janie** doesn't see her friend Meg very often since **she** moved to Winnipeg.
6. Helen told **Beth** that **she** might be late if it rains.
7. **Russ** often telephoned his brother while **he** was at camp.
8. Lynn asked her **sister** why **her** room was such a mess.
9. Our **car** hit the garage door, but **it** wasn't badly damaged.
10. **Mr. Ulanowski** told Joe that **he** should check the oil level again.

practice 3

Each of the following items contains a pronoun in heavy type with no reference. Rewrite each sentence, replacing the pronoun with more specific words.

1. Ballet is a difficult profession because **they** have to spend so much time practicing.
2. Andy twisted his ankle on the stairs, **which** kept him from going biking with us.
3. **It** says that the acting in Peter Bogdanovich's new movie is excellent.
4. We decided to take the long trail back to camp. **That** was a mistake.
5. Because of the energy crisis, **they** are looking for new sources of offshore oil.
6. **It** says in the paper that the mayor will not run for reelection.
7. Ron and his sister now take turns setting the table and washing the dishes. **This** seems to work.
8. Chin Mui found her wristwatch on the floor by her locker, **which** was lucky.
9. Mom and I went to three different plant shops but couldn't find **one** we wanted.
10. On a rainy day the children have to play indoors. **It** was really bad this morning.

← application →

Each of the following items contains a pronoun in heavy type with unclear reference or no reference. Rewrite each sentence to correct the pronoun reference.

1. Since that bad accident at the intersection, **they** are trying to have a traffic light installed.
2. Although my older brother is studying accounting, I don't want to be **one.**
3. Julie dialed the same wrong number three times, **which** was embarrassing.
4. The children followed the band members until **they** reached the railroad crossing.

5. There's a new ice-cream store on the next block. Let's get **one.**
6. Rishi usually walks Bill to school, but today **he** took the bus.
7. Martha and Beth often study together, but yesterday **she** was sick.
8. Elva told her sister that **she** was going to be late.
9. The instructions for assembling the kite were missing. **It** made the job much harder.
10. The students and the teachers are usually evenly matched in softball, but last week **they** were badly beaten.

17 *This, That, These, and Those*

The words *this, that, these,* and *those* are called "pointing words" or demonstratives. *This* and its plural *these* point to something near the speaker; *that* and its plural *those* point to something away from the speaker.

> **This book** is the one I want; I've already read **that one.**
>
> Which pictures do you want, **these** or **those?**

Notice that the demonstratives can stand alone as pronouns or they can be used as noun modifiers.

> **These** are Kevin's records. [pronoun]
>
> **These records** are Kevin's. [modifier of noun *records*]

Using demonstratives correctly is important both in speaking and in writing. Here are three common problems that you should be careful to avoid:

This Here Sometimes the words *here* and *there* are incorrectly added to *this, that, these,* and *those.* Remember that the words *this* and *these* already mean "here, near the speaker" and that the words *that* and *those* mean "there, away from the speaker." So, the words *here* and *there* are really unnecessary repetitions. Be careful not to use *here* or *there* directly after the demonstratives.

> *__This here__ is the one I mean.
>
> **This** is the one I mean.

*The asterisk (*) identifies incorrect sentences.

***That there box** is too heavy to lift.

That box is too heavy to lift.

Of course, you can use the words *here* and *there* correctly in other places in a sentence that contains a demonstrative.

Let's put **these boxes** over **there** by the door.

Them Sometimes the pronoun *them* is incorrectly used in place of *these* or *those.* Remember that *them* is the objective form of a personal pronoun. Use *them* in the object position as a pronoun.

Charlie saw **them.** [direct object]

Edie gave **them** the photographs. [indirect object]

Sue had never heard of **them.** [object of preposition *of*]

Since *them* is an objective form, you should never use it as a subject in place of *these* or *those.*

***Them** are the ones I told you about.

Those are the ones I told you about. [as subject]

Since *them* is always a pronoun, you should never use it as a modifier in place of *these* or *those.*

*Paul chose **them shirts** himself.

Paul chose **those shirts** himself. [as modifier]

These Kind Of The words *kind, sort,* and *type* are singular. Each has a regular plural form—*kinds, sorts, types.* Use the singular *this* and *that* with *kind, sort,* and *type* and the plural *these* and *those* with *kinds, sorts,* and *types.*

Mom buys **this kind** of apple for cooking. [singular]

Mom buys **these kinds** of apples for cooking. [plural]

I find **that sort** of story depressing. [singular]

I find **those sorts** of stories depressing. [plural]

Do you like **that type** of song? [singular]

Do you like **those types** of songs? [plural]

Sometimes the singular words *kind, sort,* and *type* are followed by *of* and a plural noun. Be sure you use the singular words *this* and *that* to agree with *kind, sort,* and *type,* not with the plural noun that follows *of.*

*Have you tried **these kind of cookies?** [incorrect]

Have you tried **this kind of cookies?** [correct]

practice 1

In seven of the following sentences, *here* and *there* are used incorrectly. Rewrite these sentences correctly. Three of the sentences are correct. Write *correct* after the number of each of these sentences.

1. That there student by the lunch counter is Rita's cousin.
2. Sally said that this here story is the best she's ever read.
3. These science reports should be kept right here where everyone can use them.
4. You'd better use Mrs. Collins's typewriter; that there one needs a new ribbon.
5. Why not move that plant over here near the window?
6. Just put those there crates against the garage wall.
7. That there swimmer is probably the best on the team.
8. Julian said he took two hours to do these here math problems.
9. Are there any more of those blueberry muffins left?
10. No one seems to have any idea where these here books belong.

practice 2

Rewrite each of the following sentences, choosing the correct word from the words in parentheses.

1. Did you know that __ commercials were written by a well-known songwriter? (those, them)
2. __ are the three things you must have—a parka, strong boots, and warm gloves. (Them, These)

3. Two students are still writing; ask ___ to hand in their papers now. (those, them)
4. Look! ___ must be the magazines for which Neelam was looking. (These, Them)
5. The twins were so excited about their birthday that we gave ___ one of their presents early. (these, them)
6. ___ beautiful old trees are being cut down because of Dutch elm disease. (These, Them)
7. Watching ___ dolphins perform was the best part of the show at the aquarium. (them, those)
8. The spareribs were delicious; everyone likes the way Uncle Mal cooks ___. (these, them)
9. I think both ___ basset hounds should win a prize. (those, them)
10. Now that the new bypass is complete, it's time to throw ___ old road maps away. (them, those)

practice 3

Rewrite each sentence, choosing the correct word from the words in parentheses.

1. Betsy had never seen ___ kind of animals before. (that, those)
2. It's always a pleasure to have ___ sort of campers in our national parks. (this, these)
3. Dave weakened his argument by presenting ___ kinds of half-truths. (those, that)
4. Do you like ___ type of coffee? (this, these)
5. It's often a mistake to listen to ___ type of rumors. (that, those)
6. Warm, moist weather seems to be good for ___ types of flowers. (that, those)
7. It's very important to recognize ___ sort of mushroom as poisonous. (this, these)
8. Sometimes it's hard to be on the same team with ___ kinds of players. (that, those)
9. Exact temperature control is important for ___ type of machines. (these, this)
10. Jill doesn't like to write with ___ sorts of pens. (this, these)

← application →

Rewrite the following paragraphs, removing the words *here* and *there* wherever they are not needed and changing any incorrect demonstratives.

> Arabian horses, which are noted for their beauty, stamina, and grace, have been used for hundreds of years to develop new breeds. But some admirers of these type of horse are interested in maintaining the pure breed in this here country and in the United States.
>
> A man named Homer Davenport, who first saw these kind of horses at the Chicago World's Fair in 1893, later succeeded in importing 27 of them in 1906 into the United States. One of the most famous of them Davenport imports was Buffalo Bill's horse, Muson.
>
> Today about 180 Arabian horses go back directly to the Davenport imports. Them highly prized horses are called Davenport Arabians and are found both in this here country and in the United States.

18 Subject/Verb—Agreement in Number

If a word refers to a single person or thing, it is singular in number. Words like *tree* and *glass* are singular. If a word refers to more than one person or thing, it is plural in number. Words like *trees* and *glasses* are plural.

When you write a sentence, make certain that the subject and the verb agree in number. That is, if the subject is singular, the verb should be singular. If the subject is plural, the verb should be plural. The following subjects and verbs agree in number:

Singular Subject	+	*Singular Verb*	*Plural Subject*	+	*Plural Verb*
the cat		watch**es**	the cat**s**		watch
a box		break**s**	box**es**		break
he / she / it		work**s**	they		work
you alone / I		remain	you three / we		remain

Notice that, except for pronouns, the endings *–s* and *–es* occur both with subjects and with verbs. However, there is an important difference. *Plural* subjects—but *singular* verbs—have *–s* or *–es* endings.

Notice, too, that all the verbs are present tense. Since most past-tense verbs are the same with both singular and plural subjects, they do not cause agreement problems: *Jo cried; we cried.*

Finally, notice that *you* and *I* are exceptions. They always occur with a verb that doesn't end in *–s* or *–es.*

The verb *BE* is also a special exception. It has three forms for the present tense—*am* and *is* for singular subjects and *are* for plural subjects. *BE* also has two forms for the past tense—*was* and *were.* Use the singular *was* with singular nouns and with the pronouns *I, he, she,* and *it.* Use the plural *were* with plural nouns and with *you, we,* and *they.*

Remember, adding the ending *–n't* to a verb doesn't affect the number agreement of subject and verb. Be careful to use subject-verb combinations like the singular *she doesn't* and the plural *you weren't,* not **she don't* and **you wasn't.*

Now, look at the verbs and their negatives in the two columns that follow. If you remember that a singular verb often appears with *–s* or *–es,* you will find it easier to make your subjects and verbs agree in number.

Singular Subjects and Verbs	*Plural Subjects and Verbs*
Tom does, Tom doesn't	the boys do, the boys don't
she is, she isn't	they are, they aren't
she was, she wasn't	they were, they weren't
he does, he doesn't	they do, they don't
it has, it hasn't	they have, they haven't

practice 1

Rewrite each of the following sentences, choosing the correct verb from the words in parentheses.

1. Sharon usually (rides, ride) her bike to school, (don't doesn't) she?
2. You (were, was) the one who suggested the idea. Amy (were, was) not in favor of it.

*The asterisk (*) identifies incorrect sentences.

3. Jodie (watch, watches) the tryouts every chance she (has, have).
4. If Martin (haven't, hasn't) done his homework yet, ask if he (want, wants) to go to the library with us.
5. Erika (don't, doesn't) speak any Swedish, but her grandmother (does, do).
6. My mother (take, takes) her vacation in August; we (hope, hopes) to visit my cousins in Nova Scotia.
7. Whenever Joe (have, has) a cold, he always (get, gets) a sore throat.
8. Ms. Tzylou (urges, urge) us to practise speaking French whenever we (is, are) with our French-speaking friends.
9. Jack (say, says) he (doesn't, don't) want to see that movie.
10. The band members (doesn't, don't) have enough time to practice, (do, does) they?

practice 2

Rewrite the following sentences, in which the subjects are printed in heavy type. If a subject is singular, change it to plural. If a subject is plural, change it to singular. Be sure to make the verb agree with the subject in number. Remember, the plural of *he, she,* and *it* is *they.*

1. The **glasses** break easily.
2. Aren't the **plants** dry?
3. The **watches** have broken.
4. The **winners** go first.
5. Was **she** cutting the grass?
6. **She** is painting the hall.
7. Do the **roads** need fixing?
8. The **machine** starts fast.
9. The **doors** aren't locked.
10. The **players** are on stage.
11. The **ties** match it.
12. **It** has been mended.
13. Does **she** like pizza?
14. **It** doesn't work well.
15. The **hikers** need rest.
16. The **movie** was no good.
17. Hasn't **she** called yet?
18. The **dog** wasn't barking.
19. Are the **bananas** ripe?
20. **He** doesn't want to go.

application

Rewrite the following paragraph. Correct any errors in subject-verb agreement by changing the verb to make it agree in number with the subject.

My brother Hank have a strange eating habit at breakfast. Each morning he wait until the last moment to get up. Then he have to rush to catch the bus, so he eats breakfast as he move around, picking up his books and coat. Mother keep trying to get him to sit down, but Dad say at least he is eating. And a glass of milk with a piece of toast is better than no breakfast at all. It look to me as if he don't know what he am eating or what he am missing. If he slowed down enough to see what good things was on the table, he might decide to get up early enough to enjoy them. But until then, we has a study in perpetual motion each morning at breakfast.

19 Subject/Verb — The Real Subject

To make the subject and the verb of a sentence agree in number, always match a singular subject with a singular verb and a plural subject with a plural verb. This is easy to do when the subject and the verb are in their normal order—the subject coming right before the verb.

Janie watches TV only on weekends. [singular subject and verb]

Those **puppies love** to eat. [plural subject and verb]

However, in longer, more complicated sentences, finding the real subject can be a problem. And you must know what the real subject is before you can make the verb agree.

Sometimes words other than the subject seem to control the number of the verb. When you write a sentence with *is, are, was,* or *were* as the verb, you may be tempted to make the verb agree with a noun that follows the verb.

*The best present **were the books** you gave me.

Don't make this mistake. Remember that the subject and the subject alone determines the number of the verb.

*The asterisk (*) identifies incorrect sentences.

The best **present was** the books you gave me.

These **kittens are** your problem.

Your **problem is** these kittens.

Often there is more than one subject in a sentence, and sentences with a compound subject can cause agreement problems. When you write a sentence with two subjects joined by *and,* use a plural verb.

Toby and Jim are reporting on the same book.

However, if the two subjects you join with *and* are thought of as a single unit, use a singular verb.

Ham and eggs makes a hearty breakfast.

When you write a sentence with two singular subjects joined by *or* or *nor,* use a singular verb.

Kathy or Cecile is making the announcement tomorrow.

Neither **Kathy nor Cecile wants** to do it.

When you write a sentence with a singular subject and a plural subject joined by *or* or *nor,* make the verb agree in number with the subject closer to it.

Either Sam or the **girls are** going to wash the car.

Either the girls or **Sam is** going to wash the car.

Sometimes groups of words come between the subject and the verb. Phrases beginning with expressions like *as well as, together with,* and *in addition to* are not part of the subject. Make the verb agree with the real subject, which does not include the phrase.

Max, as well as John, **has** already taken auto shop.

I think **Miss Gilbert**, in addition to Mr. Picard, **is** going to the French Club picnic.

The **players**, together with the coach, **are** going by bus.

The **coach**, as well as the players, **is** going by bus.

If you use a prepositional phrase following the subject, make the verb agree with the real subject, not with the last word in the prepositional phrase.

> The **ability** of the gymnasts **is** breathtaking.
>
> Several **windows** in the building **were** broken.

If you use a negative expression beginning with *not* or *but not* after the subject, make the verb agree with the real subject, not with the last word in the expression.

> **Betsy and Joe**, not their father, **are** weeding the garden.
>
> **Jeff**, but not his sisters, **has** agreed to baby-sit.

To test the agreement of subject and verb when they are separated by an interrupting phrase, read the sentence without the interrupting phrase.

Beware of agreement problems in sentences that contain a subordinate clause beginning with *who, which,* or *that.* When you write a sentence like this, be sure the verb in the subordinate clause agrees with the word that *who, which,* or *that* refers to.

> I just saw a **dog that catches** Frisbees in its mouth.
>
> This luggage belongs to the **swimmers who are** competing today.
>
> That's the **trail that was** closed because of a rockslide.

Sometimes the subject and the verb are not in their usual order—that is, the verb comes first: ***Was*** *Ella upset?*

When the word *here* or *there* begins a sentence, it is not the real subject but simply serves to introduce the rest of the sentence. In such sentences, the verb usually comes before the subject.

> Here **comes** the **bus**. Here **are** your test **papers**
>
> There'**s** no **milk** left. There **are** three **sandwiches**, though.

Sometimes in a sentence that begins with a prepositional

phrase, the verb comes before the subject. When you write this kind of sentence, be sure the verb agrees with the real subject.

On the shelf **were** two broken tennis **rackets.**

Among the contestants **was** my **brother** Alan.

practice 1

Rewrite each of the following sentences, drawing a line under the subject and choosing the verb that agrees in number with the subject.

1. The mystery (were, was) the three footprints outside the window.
2. The frequent thunderstorms (were, was) a new experience for us.
3. The puppet shows (is, are) always a favorite entertainment for the children.
4. The two brothers (are, is) the backbone of the organization.
5. These plants (were, was) my responsibility last month.
6. The most disagreeable part (were, was) the traffic jams.
7. Phil's favorite dessert (is, are) chocolate brownies.
8. The reason for the delay (are, is) those machines.
9. Piston rings (is, are) the seal between the piston and the cylinder wall.
10. The goal of dental hygiene (is, are) healthy gums and teeth.

practice 2

In the following sentences, two subjects are joined by *and, or,* or *nor.* In seven sentences, subjects and verbs do not agree in number. Rewrite these sentences, changing the verbs to agree with the subjects. Three sentences are correct. Write *correct* after the numbers of these sentences.

1. Neither the meat loaf nor the rice is ready yet.
2. Donna said that Helen and Judy has been practicing hard.

3. Either Dad or my brothers is making pancakes.
4. Debbie and John is planning to bring their records.
5. Joe and Nick are building a tree house in the backyard.
6. Either the bus or the subway are convenient to get to the ball park.
7. Doug or Carole is going to meet us at the bus terminal.
8. I thought Asha or her brother were going to call us.
9. The saw and the hammer was in the tool chest.
10. Neither sun nor rain damage this house paint.

practice 3

Rewrite each of the following sentences, choosing the verb that agrees in number with the subject. In sentences 1–10, test your choice by crossing out the interrupting phrase. In sentences 11–15, test your choice by drawing a circle around the word that *who* or *that* refers to.

1. Nancy, but not Sheryl, (is, are) working on the stage crew for the class play.
2. Miss Scully, together with her students, (are, is) running around the track.
3. The announcement of the awards (has, have) been scheduled for tomorrow night.
4. Chemistry, in addition to biology, (require, requires) several hours of lab work each week.
5. The bridge over the old railway tracks (was, were) being repaired last week.
6. The hot dogs, along with the potato salad, (has, have) to be kept in the refrigerator.
7. The points inside the distributor of a car usually (needs, need) changing every 20 000 kilometres.
8. Frank, as well as his sister, (don't, doesn't) like bacon.
9. The washing machine, not the dishwasher, (are, is) making a strange noise.
10. The leftovers from dinner last night (were, was) used for lunch today.
11. The students that (was, were) organizing the concert found out how expensive musical talent can be.
12. The typewriter that (give, gives) us the most trouble is the old standard one in the corner.
13. My cousins who (is, are) moving to Calgary will probably come back to visit us next summer.

14. This is the old map that (show, shows) the Indian trail.
15. There's the squirrel that (has, have) been taking food from the bird feeder.

practice 4

In each of the following sentences the verb precedes the subject. Rewrite each sentence, drawing a line under the subject and choosing the verb that agrees in number.

1. Just inside the front door (is, are) the old coatrack and the umbrella stand that Grandfather made.
2. (Have, Has) the record player come back from the shop?
3. Here (is, are) the packages I told you about.
4. (Don't, Doesn't) your brother work at that supermarket?
5. There (are, is) something important I want to tell you.
6. (Weren't, Wasn't) you going to the dentist today?
7. Suddenly around the corner (come, comes) Bobbie and Joan on a bicycle built for two.
8. By Friday afternoon there (wasn't, weren't) any tickets left.
9. Here (are, is) the strangest part of the whole story.
10. (Haven't, Hasn't) you ever read about flying saucers?

← application →

Rewrite the following paragraphs, changing any verb that does not agree with its subject in number.

Today a small number of students from various countries around the world have the opportunity to study at a United World College. One of three such colleges are located in Canada. Named after the late Prime Minister and Nobel Peace Prize winner, the Lester B. Pearson College of the Pacific are located on the shores of Pedder Bay, thirty kilometres from Victoria, B.C.

Boys or girls wishing to attend the college is required to compete for one of the 100 places that is available each year. They must have completed eleven years of academic study and be strongly recommended by their secondary school principal. Whenever possible they is also interviewed by a selection committee. The average age of applicants are 15½ to 17½. At the end of two years at the residential college, students takes the International Baccalaureate examinations that is prepared by the International Baccalaureate office in Geneva, Switzerland.

20 Subject/Verb – Determining Number

If you can't tell whether the subject is singular or plural, you'll obviously have a problem choosing a verb that agrees with it. When no general rule of agreement seems to apply, you must deal with each subject as an individual case. The specific meaning you want the subject to carry will often determine its number.

Certain nouns like *band, team, committee,* and *family* name a group and so are often called collective nouns. In most cases you will want such a noun to mean the group as a single unit. Then use a singular verb with it.

> Because of the traffic, the **band was** late in arriving for the concert. [a single unit]

> The **committee** always **meets** in this room promptly at 8:00 p.m. [a single unit]

Occasionally, you may want such a noun to mean the various members of the group. Then use a plural verb with it.

> The **band are** having their new uniforms individually tailored. [the various members]

> The **committee are** arguing among themselves about fines for lateness. [the various members]

Certain frequently used words like *each, some,* and *several* do not name definite persons or things. Such words are often called indefinite pronouns.

Some indefinite pronouns are always singular: *each, either, neither, anybody, somebody, nobody, everybody, one, no one, anyone, someone,* and *everything.* Use a singular verb when you write a sentence with one of these singular words as the subject.

> **Everyone is** ready. **Nobody cares. Someone has** been here.

Remember that a prepositional phrase following the subject does not affect the agreement of subject and verb.

> **Neither** of the dancers **was** present.
>
> **Each** of the students **has** received a certificate.
>
> **Everything** in the serving bowls **is** hot.

Some indefinite pronouns are always plural: *several, many, both,* and *few.* Use a plural verb when you write a sentence with one of these plural words as the subject.

> **Many are** homeless. **Few were** able to escape.
>
> **Several have** already left. **Both** of us **were** waiting.

Other words like *none, some, any, most, part,* and *all* and fractions like *half* and *two thirds* may be either singular or plural, depending on how you use them. Before you write a sentence with one of these words as the subject, decide whether you want the word to tell *how much* or *how many.* If you want your sentence to answer the question *how much,* use a singular verb.

> **None** of the floor **was** refinished. [How much of the floor **was** refinished?]
>
> **Two thirds** of the house **is** already painted. [How much of the house **is** already painted?]

If you want your sentence to answer the question *how many,* use a plural verb.

> **None** of the floors **were** refinished. [How many of the floors **were** refinished?]
>
> **Two thirds** of the houses **are** already painted. [How many of the houses **are** already painted?]

Some expressions referring to measurement or to amounts of time or money may be either singular or plural. When you want such an expression to mean items grouped together to form a single unit, use a singular verb with it.

Three eggs is a lot for one cake. [a single unit]

Two weeks seems a long time to wait. [a single unit]

When you want such an expression to mean several separate items, use a plural verb with it.

Three eggs are rolling off the table! [several items]

The final **two weeks are** set aside for review and testing. [several items]

The word *number* can also be either singular or plural. When you write ***the*** *number,* it means a single unit, the total amount, and you should use a singular verb.

The number of citizens signing the petition **is** growing steadily.

When you write ***a*** *number,* it means "several, more than one," and you should use a plural verb.

A number of citizens **are** concerned about air pollution.

Some nouns look plural because they end in *–s.* When you use a noun like *mumps, news, measles,* or *physics* in a singular sense, use a singular verb with it.

Social Studies is usually taught by Ms. Cohen and Mr. Mitchell.

The **news** about this year's wheat crop **has** been good.

Some words like *scissors, trousers,* and *eyeglasses* are always plural, even though they name a single object.

The **scissors are** never where they belong.

But if you add the words *a pair of,* use a singular verb.

A pair of scissors is essential for sewing.

Use a singular verb with the title of a book, play, poem,

movie, or television program, even though the words in the title may be plural.

> ***The Guns of August* is** a historical account of the months leading up to World War I.

practice 1

Rewrite each of the following sentences, choosing a verb from the words in parentheses. Consider the subject in sentences 2, 4, 5, 7, 8, 9, and 10 to be a single unit.

1. The crew of fire fighters (was, were) aided by hundreds of tourists who were vacationing in the immediate area of the fire.
2. Our community (are, is) trying to promote the use of bicycle lanes in the main business area.
3. The graduating class (hasn't, haven't) planned their spring dance yet.
4. (Aren't, Isn't) the legislature in session today?
5. The audience (was, were) quiet as the performance began.
6. (Wasn't, Weren't) the panel of athletic directors going to present their personal views on television?
7. The jury (was, were) prompt in reaching its decision.
8. The swimming team (is, are) eager to defend its title.
9. The Outdoor Club (don't, doesn't) plan to increase its dues this year.
10. The coast guard (patrol, patrols) rivers as well as the seacoasts.

practice 2

In seven of the following sentences, verbs do not agree with subjects. Rewrite these sentences, changing the verbs to agree with the subjects. Three sentences are correct. Write *correct* after the numbers of these sentences.

1. Everyone in the tour buses are eager to get started.
2. Does either of you have a quarter I could borrow?
3. Everything happen to me.

4. Somebody have forgotten to turn off the car lights.
5. Neither of these flashlights works.
6. Have anybody seen my history book?
7. Many in the theater was upset by the movie.
8. Several of the new dishes has already been broken.
9. Both my sisters are taking biology this year.
10. Few was hurt by the train's sudden stop.

practice 3

Decide whether the subject in each sentence should be singular or plural by asking the questions *how much* and *how many*. Then rewrite each sentence, choosing a verb from the words in parentheses.

1. Fortunately, none of the ice cream (has, have) melted.
2. Most of the books on this reading list (are, is) paperbacks.
3. Part of the trouble (was, were) the broken handle.
4. Half the students in the grade nine class already (knows, know) how to swim.
5. All the water in the radiator (has, have) leaked out.
6. None of this explanation (makes, make) any sense to me.
7. (Is, Are) any of the iced tea left?
8. All the papers left near the open window (were, was) damaged by the rain.
9. Three fourths of the money (have, has) been set aside for paying the band.
10. Some of the oil (has, have) spilled out onto the floor.

practice 4

Rewrite each of the following sentences, choosing the verb from the words in parentheses.

1. The number of students studying French (is, are) increasing each year.
2. *King Solomon's Mines* (remain, remains) one of my favorite old movies.

3. Fifteen pennies (were, was) lying on top of the pile of newspapers. [Consider the subject to be several items.]
4. Do you think that measles (are, is) contagious?
5. A pair of eyeglasses (has, have) been turned in to the office.
6. Thirteen dollars (seems, seem) a lot to pay for a head scarf. [Consider the subject to be a single unit.]
7. A number of students (has, have) asked about a course in horticulture.
8. These trousers (look, looks) too baggy.
9. Forty-five minutes (aren't, isn't) long enough to study for a history test. [Consider the subject to be a single unit.]
10. Mathematics (have, has) always been Ron's best subject.

application

Write sentences of your own, using the following words as subjects and choosing a verb from the words in parentheses. Consider the subject in sentence 6 and in sentence 9 as a single unit.

1. Two thirds of the lemonade (is, are) ___.
2. The number of letters (was, were) ___.
3. Each of the players (need, needs) ___.
4. Most of the building (remains, remain) ___.
5. Many of the test papers (show, shows) ___.
6. Three weeks (are, is) ___.
7. A number of teachers (require, requires) ___.
8. The news (have, has) been ___.
9. Our volleyball team (is, are) ___.
10. *Alice's Adventures in Wonderland* (are, is) ___.

21 Forms of *BE*

When you write, be sure you include *BE* whenever it is needed to complete a sentence. Be sure also that you use the correct form of *BE.*

The first sentence in each of the following pairs is faulty, either because a *BE* verb is omitted entirely or because a wrong form of *BE* is used. The second sentence in each pair illustrates the correct use of *BE* in writing.

*The Browns ↓ coming here this afternoon. [omitted]
The Browns **are** coming here this afternoon.

*Dodie ↓ the smartest girl in class. [omitted]
Dodie **is** the smartest girl in class.

*They **be** late all the time. [wrong form]
They **are** late all the time.

*We **was** just watching the team practice. [wrong form]
We **were** just watching the team practice.

Remember that *BE* has three forms in the present tense—*am* and *is* for singular subjects and *are* for plural subjects and for the pronoun *you. BE* also has two forms in the past tense—*was* for singular subjects and *were* for plural subjects and for the pronoun *you.*

Use a form of *BE* as the helping verb that occurs with *-ing* verbs.

I **am leaving** now. I **was** just **finishing** when you called.

Gladys **is going**, too. She **was hoping** to go yesterday.

You **are taking** a long time. You **were sleeping** when we arrived.

Use a form of *BE* as the main verb to link nouns or modifiers with the subject.

Willis **is** optimistic. The basketball players **are** agile.

Janet **is** my cousin. Al and Doug **are** the class clowns.

The camera **is** on the table. The flash cubes **are** in the drawer.

When you write informally, you may want to use a contraction of *BE* or of a negative used with *BE.* Be sure to show contractions in your writing by using an apostrophe to indicate the missing letters.

He's tired. **I'm** listening. **We're** almost ready. **They're** here.

They aren't here. **They're not** here. **It's not** broken. **It isn't** broken. **I'm not** hungry.

*The asterisk (*) identifies incorrect sentences.

Aren't you ready yet? **Isn't she** your cousin? I'm on time, **aren't I?**

Remember that *'re* is a contraction of *are,* not *were,* and that *aren't* is used in negative questions with *I.*

practice 1

Each of the following sentences is written correctly. Rewrite each sentence. If the verb form in heavy type is in the present tense, change it to the past tense; if it is in the past tense, change it to the present tense.

1. Joe and Dee **were** watching *The National News* on television.
2. I **wasn't** the team captain.
3. You **are** the first one to arrive.
4. Everyone **is** talking at once.
5. Jeff and Ashok **were** upset about the extra rehearsal.
6. The apples **are** in the refrigerator.
7. Jake **is** working at the garage after school.
8. **Weren't** you surprised at the news?
9. **I'm** not going to say anything.
10. We **aren't** the last to finish.
11. Sharon's mother **was** a computer programmer.
12. Why **isn't** the record player running?
13. What **were** you planning to do on Saturday?
14. **Aren't** I supposed to give my report today?
15. **Was** Linda at the movies this afternoon?

practice 2

Rewrite each of the following sentences, adding the appropriate form of *BE.* Use the present or past tense as indicated by the word in parentheses at the end of each sentence. Be sure you use a singular form with a singular subject and a plural form with a plural subject.

1. We ___ trying to telephone you. (past)
2. Tony ___ eating lunch when we got there. (past)
3. Ben ___ president of the Outdoor Club last year. (past)
4. Diane ___ baby-sitting tonight. (present)
5. ___ Marilyn taking the bus or the train? (present)

6. Our class ___ the first to use the new pool. (past)
7. Who ___ you going to talk to? (past)
8. My bicycle ___ that red and white one. (present)
9. What ___ we having for dinner tonight? (present)
10. ___ I supposed to let you know the address? (past)
11. We ___ standing in the ticket line by six o'clock. (past)
12. ___ the door open when you got there? (past)
13. I ___ really upset by his attitude. (present)
14. These keys ___ on the sidewalk. (past)
15. ___ Ruth inside the supermarket? (past)
16. ___ we going to meet your brother? (present)
17. Our next-door neighbors ___ away this week. (present)
18. ___ you pleased about the final score? (present)
19. Nobody ___ perfect. (present)
20. Why ___ you running toward the gym yesterday? (past)

← application →

Rewrite the following paragraphs, adding the correct form of *BE* wherever you see the words *past* or *present* in parentheses. Use these words to choose the correct tense. Make certain also to choose singular verbs for singular subjects and plural verbs for plural subjects.

> New guidelines (present) in effect that eliminate separate physical-education classes for boys and girls, except in contact sports. Although schools (present) given three years to comply, many coed classes (present) already the routine at our high school.
>
> Instructors and students report that the classes (present) going very well. Many students say that the coed classes (present) more fun. Contact sports like soccer, basketball, softball, and field hockey (present) still taught separately, but swimming, volleyball, tennis, and badminton (present) mixed.
>
> When gym classes (past) always separate, there (past) much duplication of teaching, and many students (past) made to take activities they didn't like. Now, with mixed classes, everyone (present) benefiting from a more varied program, and many students (present) showing an increased interest in physical education.

22 Irregular Verbs

When you write, you must use different forms of a verb to indicate different times or tenses.

I **walk** here every day. [present or simple form]

I am **walking** here right now. [*–ing* form]

I **walked** here yesterday. [past form]

I have **walked** here often. [past participle form]

To choose the correct verb form to use in a sentence, you need to know three of the verb's four basic forms—the simple form, the past form, and the past participle form. These three forms are called the principal parts of a verb. The fourth form—the *–ing* form—is always made the same way, by adding *–ing* to the simple form.

Verbs in English are either regular or irregular, depending on how their principal parts are formed. Most verbs are regular. That is, you regularly add *–ed* or *–d* to their simple forms to make both the past and past participle forms. Here are the principal parts of two regular verbs:

Simple	*Past*	*Past Participle*
look	look**ed**	(have) look**ed**
race	race**d**	(have) race**d**

As you can see, learning the principal parts of regular verbs is no problem because they are made in a completely regular way.

Most problems happen with irregular verbs. As their name suggests, there is no regular way to make the past and the past participle forms of these verbs.

Some irregular verbs require a vowel change to make their past and past participle forms.

Simple	*Past*	*Past Participle*
beg**i**n	beg**a**n	(have) beg**u**n
f**i**nd	f**ou**nd	(have) f**ou**nd

Other irregular verbs also require an *–en* or *–n* ending for the past participle.

Simple	*Past*	*Past Participle*
fall	fell	(have) fall**en**
take	took	(have) take**n**

Still other irregular verbs require a completely different word for each form.

Simple	*Past*	*Past Participle*
go	went	(have) gone
am, is, are	was, were	(have) been

Finally, a few irregular verbs, such as *burst, cost, cut, hit, let,* and *put,* require no change at all.

Be careful not to confuse the past and past participle forms of irregular verbs. Never use a helping verb with the past form.

Dave **broke** the record for the 100-yard dash. [past form]

However, always use a helping verb with the past participle form. You may use a form of *HAVE—have, has, had—* or a form of *BE—am, is, are, was, were, being, been—*or both.

Dave **had broken** another track record six months before this track meet. [*HAVE* + past participle]

The provincial high-school record for the 100-yard dash **was broken** by Dave. [*BE* + past participle]

Another track record **has been broken** by Dave since that track meet. [*HAVE* + *BE* + past participle]

Here is a list of the principal parts of some frequently used irregular verbs. Use this list for review and reference as you do the practice exercises that follow.

Simple	*Past*	*Past Participle*
beat	beat	(have) beaten
become	became	(have) become
begin	began	(have) begun
bite	bit	(have) bitten
blow	blew	(have) blown
break	broke	(have) broken
bring	brought	(have) brought
buy	bought	(have) bought
catch	caught	(have) caught

Simple	*Past*	*Past Participle*
choose	chose	(have) chosen
come	came	(have) come
do	did	(have) done
draw	drew	(have) drawn
drink	drank	(have) drunk
drive	drove	(have) driven
drown	drowned	(have) drowned
eat	ate	(have) eaten
fall	fell	(have) fallen
fight	fought	(have) fought
find	found	(have) found
fly	flew	(have) flown
forget	forgot	(have) forgotten, forgot
freeze	froze	(have) frozen
get	got	(have) got
give	gave	(have) given
go	went	(have) gone
grow	grew	(have) grown
hear	heard	(have) heard
keep	kept	(have) kept
know	knew	(have) known
lose	lost	(have) lost
ride	rode	(have) ridden
ring	rang	(have) rung
run	ran	(have) run
see	saw	(have) seen
shrink	shrank, shrunk	(have) shrunk, shrunken
sing	sang	(have) sung
sink	sank	(have) sunk
speak	spoke	(have) spoken
spring	sprang, sprung	(have) sprung
steal	stole	(have) stolen
stick	stuck	(have) stuck
swim	swam	(have) swum
swing	swung	(have) swung
take	took	(have) taken
teach	taught	(have) taught
tear	tore	(have) torn
throw	threw	(have) thrown
wear	wore	(have) worn
win	won	(have) won
write	wrote	(have) written

practice 1

Each of the following sentences is written correctly. Using the chart of irregular verbs, identify each verb form in heavy type as either *simple, past,* or *past participle.* Write the correct label for the verb form after the number of each sentence.

1. Sue **wrote** home once a week when she was at camp.
2. We sometimes **choose** vanilla ice cream just for a change.
3. Jeanne **tore** her coat on the door handle yesterday.
4. Haven't you ever **seen** the giraffes at the zoo?
5. When you're ready, **begin** to write.
6. Ted **caught** the last bus leaving for the city.
7. Ah Chong and Ken Ying often **buy** lunch at the cafeteria.
8. Had the bell **rung** by the time you arrived?
9. I **took** this photograph two years ago.
10. Uncle Henry has **driven** to Saskatchewan for his vacation.

practice 2

Rewrite each of the following sentences, changing the verb in heavy type from the simple form to the past form.

1. Mr. and Mrs. Reilly **grow** vegetables in the community lot.
2. I **know** about your plans for a surprise party.
3. Mom and Dad **fight** rush-hour traffic each night.
4. We usually **drink** milk at suppertime.
5. Each summer, my cousins from Manitoba **come** to visit.
6. Carmen and Patty **ride** the bus instead of the train.
7. I **bring** my math problems home for Aunt Julie to read.
8. The instructors **teach** diving as well as swimming.
9. Carol and Dick **wear** blue jeans as often as possible.
10. In hot weather, the campers **swim** twice a day.
11. The gears **stick** in very cold weather.
12. The Andersons often **drive** us to the shopping center.
13. The children always **blow** out the candles on the birthday cake.

14. Carla and Ron **eat** pizza for lunch at least once a week.
15. I seldom **win** at Scrabble.

practice 3

Rewrite each of the following sentences, adding the form of *HAVE* in parentheses and changing the verb in heavy type from the past form to the past participle form.

1. Elva **flew** to Manila to visit her sister. (has)
2. The class **gave** a contribution to the new library. (has)
3. Gretchen **threw** the newspaper away. (had)
4. The union leaders **spoke** to the reporters. (have)
5. Phil **became** upset about working overtime. (had)
6. Maria and I **began** to study the guitar. (have)
7. Mom **went** to work. (has)
8. Kay and Al **drew** a sketch of their apartment. (have)
9. The surface of the pond **froze** suddenly. (had)
10. We **chose** three numbers for the next concert. (have)
11. Our English class **wrote** a television script. (has)
12. Two small boats **sank** in the harbor. (had)
13. We **beat** the Hamilton soccer team again. (have)
14. Pat **swam** the length of the pool underwater. (has)
15. That movie **ran** for two months. (has)

practice 4

Rewrite each of the following sentences, adding the past participle form of the verb in parentheses.

1. French is ___ every morning at ten. (teach)
2. Two bikes were ___ last week outside the gym. (steal)
3. Do you know where this camera was ___? (buy)
4. The national anthem is ___ before the game. (sing)
5. We were badly ___ by mosquitoes at the picnic. (bite)
6. The seats near the front have already been ___. (take)
7. I don't know how my jeans were ___. (tear)
8. These plants were ___ under artificial light. (grow)
9. Was anyone ___ in the accident? (hurt)
10. This horse is always ___ by the new students. (ride)

11. The hot dogs were ___ by the time we got there. (eat)
12. The battle against the common cold is still being ___. (fight)
13. The supplies were ___ to the isolated settlers. (fly)
14. You have been ___ to speak at the assembly. (choose)
15. After thawing for two hours, the meat was still ___. (freeze)

← application →

Rewrite the following paragraphs, adding the appropriate form—past or past participle—of each verb in parentheses.

> Early in the 1940s, zoos (**become**) increasingly concerned about the preservation or rare and endangered species, and they (**begin**) to develop breeding programs for their animals in captivity. They (**hope**) in this way to preserve some species which were (**know**) to be facing extinction in their natural habitat.
>
> In the past, zoos might have (**keep**) one animal of a rare species. They now may have (**buy**) several, with the aim of developing breeding herds. Animals may also be (**give**) or (**trade**) from zoo to zoo, often to form breeding pairs.
>
> Medical facilities at many zoos have (**grow**) to include nurseries providing special care for newborn animals. The success of this program can be (**see**) in the recent birth in captivity of pandas, gorillas, a sea otter, and a beluga whale.

23 Forms of *HAVE*

Because the verb *HAVE* functions both as a main verb and as a helping verb and is used often in special expressions, you could have difficulty choosing the proper form of *HAVE* when you write.

1. You may use a form of *HAVE* as a main verb.

Sally **has** a problem she can't solve. [present tense]

He **has** a very bad case of the flu. [present tense]

I **have** some good pictures of the picnic. [present tense]

They **have** two weeks before the deadline. [present tense]

The Joneses **have** three Angora kittens. [present tense]

We **had** company the last two weekends. [past tense]

Teri **had** nothing to say. [past tense]

Notice that in the present tense the form *has* is used with singular subjects like *Sally* and with the pronouns *he, she,* and *it.* But the form *have* is used with plural subjects like *the Joneses* and with the pronouns *I, you, we,* and *they.* Notice too that the form *had* is the only past tense form. *Had* is used with both singular and plural subjects and with all pronouns.

2. You may also use a form of *HAVE* as a helping verb.

I **have** often **wondered** about life on other planets.

Rita **has** never **mentioned** it to me.

Tony **had heard** the news before we did.

Notice that you use *had* as the helping verb when you express the earlier of two past actions. Avoid using *would have* after *if* in sentences like this one:

***If** I **would have known** that yesterday morning, I could have helped you.

If I **had known** that yesterday [earlier past time], I could have helped you [past time].

When you express a wish about the past, use *had* as the helping verb after the word *wish.* Avoid using *would have* after *wish* in sentences like this one:

*I **wish** I **would have known** that last year.

I **wish** I **had known** that last year.

3. You may use the expressions *have to, has to,* or *had to* to show necessity.

I **have to** study tonight, so I can't go along.

Fritz **had to** be back home by ten o'clock.

*The asterisk (*) identifies incorrect sentences.

4. When you speak or write informally, you may add the word *got* to *have* or *has* for emphasis.

Sally **has got** a real problem.

I **have got** to study tonight.

5. In speaking, *have* and *has* are sometimes contracted or even omitted. However, when you write, be sure you always include *have* or *has* or their contractions *'ve* or *'s.*

*He ↓ got lots of time.
He **has** got lots of time.
He**'s** got lots of time.

*You ↓ got to be quiet.
You **have** got to be quiet.
You**'ve** got to be quiet.

*She ↓ been waiting for you.
She **has** been waiting for you.
She**'s** been waiting for you.

6. The contraction *'ve* is often pronounced exactly like *of.* Don't let this pronunciation fool you into writing *of* in place of *have* or *'ve.*

*It must **of** broken.
It must **have** broken.
It must**'ve** broken.

*He should **of** called us.
He should **have** called us.
He should**'ve** called us.

practice 1

Rewrite each of the following sentences, adding *have* or *has* to replace the blank.

1. I ___ never understood that problem.
2. Diane ___ something to say to you.
3. Professional soccer ___ been played there since 1885.
4. Who ___ time to explain this to me?
5. You ___ to swim the length of the pool and back.
6. ___ anyone seen my social studies book?

7. Perhaps we __ misunderstood Mrs. Sellers.
8. Those plants __ flourished in the warm, wet weather.
9. Did you know that sweet potatoes __ high energy value?
10. Kevin __ to go to the dentist tomorrow.

practice 2

Each of the following sentences is written correctly with *had.* Rewrite each sentence, changing *had* to *has* or *have.*

1. Meena **had** a lot to tell us about her vacation.
2. The factory workers **had** a new contract.
3. The watchman **had** checked all the windows and doors.
4. Ben **had** already seen that movie twice.
5. We **had** to redo all the posters, changing the location.
6. At camp, we **had** to set the tables and sweep the floor.
7. Shirley **had** forgotten to leave the key for us.
8. Ching Sue **had** a new job at the supermarket on weekends.
9. Consumers **had** developed a new interest in small cars.
10. Max **had** rewritten his report three times.

practice 3

The following sentences express the earlier of two past actions after *if* or state a wish about the past. Rewrite each sentence, adding the helping verb *had* and the appropriate form of the main verb in parentheses.

1. Graduates sometimes wish that they __ other courses in high school. (take)
2. If Mom __ about the traffic jam earlier, she could have avoided the expressway. (hear)
3. The strawberry shortcake looks so good that I wish I __ it instead of ice cream. (order)
4. Jody wishes that she __ for track last year. (try out)
5. If Steve __ less nervous, he might have done better on the exam. (be)

6. I would have telephoned you last night if I __ your number. (know)
7. I think Linda wishes she __ that record at the sale yesterday. (buy)
8. If you __ ten more minutes, you could have met the premier. (wait)
9. Our farm would have been flooded if the river __ another two inches. (rise)
10. I wish you __ it to me at the very beginning. (explain)

practice 4

Sentences 1–10 are written correctly. Rewrite, replacing the contraction with the full form *has* or *have.* Sentences 11–15 are written *in*correctly with the word *of.* Rewrite, changing the word *of* to the correct form of *HAVE.*

1. We've been waiting for the bus for 20 minutes.
2. They've got to be ready to leave at noon.
3. You've got plenty of time to eat lunch first.
4. I've been listening to you say that for weeks.
5. My science report's got three pages of charts.
6. Kathy's been doing algebra problems for over an hour.
7. Please tell Mrs. Girard that we've got to leave now.
8. Jack's been trying to talk to you all afternoon.
9. Carl's got some new records he wants us to hear.
10. We've got to move those bikes out of the garage.
11. Mr. Pieroni ought to of explained it better.
12. You could of walked home with us yesterday.
13. Somebody must of been using the tape recorder.
14. The mail should of been here by now.
15. They might of been trying to telephone you.

← application →

Write sentences of your own, using the expressions given below. In sentences 1–10, use *has* or *have.* In sentences 11–15, use *had.*

1. Sharon ___ to leave
2. we ___ been watching
3. you ___ to listen
4. they ___ got to finish
5. it ___ always worked
6. Leo ___ a part-time job
7. they must ___ been late
8. she ___ lots of patience
9. the motor ___ been on
10. we ___ got enough space
11. if she ___ known
12. I wish I ___ understood
13. if someone ___ explained
14. Carl wishes he ___ gone
15. if Leroy ___ tried

24 Forms of *DO*

DO may be a main verb. Or, you may use it to form questions and negatives, to show emphasis, to substitute for another verb already mentioned, and to form a short question at the end of a statement. Because of its many uses, choosing the correct form of *DO* is sometimes a problem.

Ben **does** his biology homework in the lab. [main verb]

We usually **do** the dinner dishes together. [main verb]

Does it matter which one I choose? [question]

When **do** Julie and Ann practice basketball? [question]

When **does** Julie practice basketball? [question]

She **doesn't** want to hear about it. [negative]

They **don't** expect to go. [negative]

I ***do*** understand what you mean. [emphasis]

He ***does*** study! [emphasis]

Midori likes waffles as much as you **do.** [substitute]

Dan and Al don't look angry, **do** they? [short question]

Notice that the present tense form *does* is used with singular subjects and with the pronouns *he, she,* and *it.* The present tense form *do* is used with plural subjects and with the pronouns *I, you, we,* and *they.*

Remember that the negative form of *do* is *don't* and the negative form of *does* is *doesn't.* Use *doesn't* wherever you would use *does*—with singular subjects and the pronouns *he, she,* and *it.* Be careful not to mix up *don't* and *doesn't.*

*She **don't** like to miss volleyball practice.

She **doesn't** like to miss volleyball practice.

*The asterisk (*) identifies incorrect sentences.

Especially when you write, be careful not to use *don't* where you should use *doesn't!*

Why **doesn't Ginnie** ride her bike to school?

We don't use that book in class anymore.

The motor doesn't usually make that strange noise.

They don't understand our views on ecology.

Doesn't Jeff have band practice this afternoon?

I don't have to give my report tomorrow.

Don't you care about the election results?

The past tense form of *DO* is *did.* Because *did* is used with both singular and plural subjects, it usually causes no problems. However, it is sometimes confused with the past participle form *done.*

*We **done** it all by ourselves.

We **did** it all by ourselves.

To avoid this confusion, remember that *done* must always be used with a helping word like *have, has,* or *had.*

The grade eleven students **did** all the decorating for the dance.

The grade eleven students **have done** all the decorating for the dance.

practice 1

Rewrite each of the following sentences, choosing *do* or *does* to complete each sentence correctly.

1. Hélène certainly enjoys speaking French more than her brother ___.
2. The twins ___ the dishes each night after dinner.
3. Dick likes listening to Gladys Knight and the Pips as much as we ___.
4. When ___ the bus leave for the swim meet?
5. Diane and Joe often ___ their math assignment together.
6. She ___ try hard in social studies!
7. ___ Pete want to go to the movies with us?
8. Jon practices the guitar more often than I ___.
9. Where ___ this tape recorder belong?
10. ___ you think you can be ready by six?
11. Our coach always ___ her best to help us improve.
12. No one else makes pizza the way Joe ___.

13. ___ you mind if I borrow your bike?
14. Why ___ this car start in neutral instead of park?
15. I ___ get lots of exercise!

practice 2

Rewrite each of the following sentences, choosing *don't* or *doesn't* to complete each sentence correctly.

1. Jan skates very well, ___ she?
2. ___ Mr. Thompson always lock the doors at five?
3. Why ___ Gus and Al look at the secondhand motorcycle?
4. ___ he want to try out for the team?
5. Mom ___ get home until six o'clock on Thursday.
6. Bev says she ___ like broccoli.
7. ___ she usually do her homework in the library?
8. He ___ like interruptions when he's talking.
9. ___ this typewriter need a special kind of ribbon?
10. These articles ___ have the information I'm looking for.
11. Suzie already has her driver's license, but Bill ___.
12. Why ___ he listen to both sides of the question?
13. Dave likes to cook, ___ he?
14. ___ all this rain depress you?
15. Rekha ___ expect to be back before midnight.

practice 3

Rewrite each of the following sentences, choosing *did* or *done* to complete each sentence correctly.

1. What have you ___ with my old Elvis Presley records?
2. Sally said she ___ ten push-ups in gym this morning.
3. I think Dad has already ___ the dishes.
4. Whoever ___ this had better speak up soon.
5. The Feldmans ___ everything possible to make us happy.
6. Aunt Ruth ___ all the driving on the trip home.
7. We ___ our best to persuade Jack to stay for the movie.
8. The entertainment committee has already ___ the planning for the variety night.

9. Last night Russ and Barbie __ the rest of the posters for the track meet.
10. I'm sure you __ what you thought was right.

application

Rewrite the following conversation, choosing the form of *DO* from the words in parentheses to complete the sentences correctly.

> "Pete saw you running with your dog Lady yesterday. (Do, Does) she run with you every day?"
>
> "Yes, she usually (do, does). I think it's (did, done) us both a lot of good."
>
> "Did she run with you last winter, too?"
>
> "Yes, but not as often. If it's too cold or wet out, she (don't, doesn't) like to run any more than I (do, does). Has Pete (did, done) much running?"
>
> "He (did, done) a lot a few years ago, but he hasn't (did, done) any recently."
>
> "Then why (don't, doesn't) he join us some morning? It (don't, doesn't) take long, and Lady really (do, does) like it."

25 Indirect Quotations

When you write down someone's words exactly as they were spoken, you use quotation marks to show that you've written a direct quotation.

> Roy asked us, "Do you like pizza?"

When you write down someone's words without quoting them exactly, you don't use quotation marks because you are writing an indirect quotation.

> Roy asked us if we liked pizza.

However, there are other important differences in writing direct and indirect quotations besides the use of quotation marks. Being aware of these differences can help you to avoid writing an awkward indirect quotation like this one:

*Roy asked us did we like pizza.

What follows is an explanation of the differences between direct and indirect quotations. Indirect statements and indirect questions are explained.

Statements Notice the differences in the same statement written as a direct quotation and as an indirect quotation.

Jo said, **"I can't do it alone."** [direct]

Jo said **that she couldn't do it alone.**
Jo said **she couldn't do it alone.** } [indirect]

In the statement as an indirect quotation

1. There are no quotation marks.
2. The pronoun *I* shifts to *she.*
3. The verb *can't* shifts to its past tense form *couldn't* to agree with the past tense form *said.*
4. The word *that* may be added to introduce an indirect statement; however, the word *that* is sometimes omitted.

***Yes/No* Questions** Some questions can be answered by *yes* or *no.* Notice the differences in the same *yes/no* question written as a direct and as an indirect quotation.

Roy asked us, **"Do you like pizza?"** [direct]

Roy asked us **if we liked pizza.**
Roy asked us **whether we liked pizza.**
Roy asked us **whether or not we liked pizza.** } [indirect]

In the *yes/no* question as an indirect quotation

1. There are no quotation marks, and a period replaces the question mark.
2. The pronoun *you* shifts to *we.*
3. The verb *like* shifts to its past tense form *liked* to agree with the past tense form *asked.*
4. The words *if, whether,* or *whether or not* must be added to replace *do* and to introduce the indirect question.

Question-Word Questions Some questions begin with a question word like *how, why, when, where, what, which.* Notice the differences in the same question-word question written as a direct and as an indirect quotation.

*The asterisk (*) identifies incorrect sentences.

I asked them, **"When will you be ready?"** [direct]

I asked them **when they would be ready.** [indirect]

In the question-word question as an indirect quotation

1. There are no quotation marks, and a period replaces the question mark.

2. The pronoun *you* shifts to *they* and moves to a position directly after the question word *when.*

3. The verb *will* shifts to its past tense form *would* to agree with the past tense form *asked.*

practice 1

The following sentences contain statements written as direct quotations. Rewrite each sentence, adding the word *that* and changing the statement to an indirect quotation.

1. Mr. Smith announced, "The test tomorrow will take forty-five minutes."
2. Toby declared, "This song is sure to be a hit."
3. Mike said, "Ann will be upset about the flat tire."
4. Wendy said, "I've already answered the questionnaire."
5. Mr. Burton reminded us, "I expect everyone to be there by eight."
6. Joe promised his brother, "I won't do it again."
7. Dad remarked, "We have been waiting for an hour."
8. Koji said firmly, "I like the blue one best."
9. Beth whispered, "The microphone may be broken."
10. Carmen told us, "This is the tenth pollution alert this year."

practice 2

Each of the following sentences contains a *yes/no* question written as a direct quotation. Rewrite each sentence, changing the direct quotation to an indirect quotation.

1. Ms. Wronski asked, "Do you want to study horticulture?"
2. Dave asked, "Does Joe have the math assignment?"
3. Paul asked, "Do you know Ellie's number?"
4. Mary asked me, "Are you going to take the bus today?"

5. Tim asked, "Can someone tell me the right time?"
6. Jeff asked him, "Is your watch broken?"
7. Mr. Berman asked, "Has anyone finished writing?"
8. Lee asked us, "Are you planning to attend the rally?"
9. Jill asked us, "Will you buy the food for the party?"
10. Sam asked, "Do they understand the problem?"

practice 3

Each of the following sentences contains a question-word question written as a direct quotation. Rewrite each sentence, changing the direct quotation to an indirect quotation.

1. Dad asked us, "Where are you going?"
2. Mom asked us, "When will you be home?"
3. The stranger asked, "Where is the bus terminal?"
4. My brother asked, "Why does it always rain on Saturday?"
5. Carol asked, "How can I arrange to leave early?"
6. The reporter asked the premier, "What do you plan to do about water pollution?"
7. The librarian asked the children, "Which story do you want to hear?"
8. The coach asked us, "Why are you late?"
9. I asked the mechanic, "What does that noise mean?"
10. Leo asked, "Where do those hockey sticks belong?"

← application →

Rewrite each of the following sentences, changing the direct quotation to an indirect quotation.

1. Betsy asked me, "Have you seen the new dolphins at the aquarium?"
2. I told her, "I haven't, but I've read about them."
3. She explained, "I'm taking my brother on Saturday."
4. Then she asked me, "Do you want to go with us?"
5. I asked her, "What time are you going?"
6. Then I asked, "How long will you be gone?"

7. I reminded her, "My part-time job starts at five."
8. She told me, "We can go right after lunch."
9. Then she added, "We'll be back before five."
10. I told her, "That sounds fine."

26 Subjunctive Verb Forms

English verbs have three moods. When you write ordinary statements and questions, the verbs you use are in the indicative mood.

Jody **likes** to read detective thrillers.

Does she really?

When you write requests or commands, the verbs you use are in the imperative mood.

Close that door.

Please **let** him stay.

When you write certain other kinds of requests, statements contrary to fact, and wishes, the verbs you use are in the subjunctive mood.

The nurse recommended that Joe **see** a dentist at once. [not *sees*]

I wish Meg **were** here to explain it to you. [not *was*]

You can see that verb forms in the subjunctive mood differ from those you use for ordinary statements. For example, forms normally ending in *–s* in the present tense appear without the *–s,* and *were* is used instead of *was.* These special forms are usually found in writing, but they may be used in speaking as well. Here are three situations in which you should use subjunctive verb forms in your writing.

***That* Clauses** In *that* clauses after verbs like *suggest, insist, recommend, demand,* and *ask,* use the simple form of the verb in the present tense, not the *–s* form.

We **suggest that** a member of the graduating class **speak** at the sports banquet. [not *speaks*]

They **asked that** the library **remain** open until ten-thirty. [not *remains*]

The doctor **insisted that** Ben **stay** at home for two more days. [not *stays*]

Statements Contrary to Fact You often use the words *if, as if,* and *as though* to express something that is unreal or contrary to fact. That is, what you express is not true at the moment and not likely to become true in the future. When the verb in such a contrary-to-fact statement is *BE,* use the form *were* with all subjects, singular and plural.

If my brother were here now, we would have a family reunion. [My brother isn't here now.]

I was so happy, I felt **as though I were** floating on air. [I'm not floating on air.]

They treated Jan **as if she were** a porcelain doll. [Jan isn't a porcelain doll.]

Wishes You use the word *wish* to express something not true that you want to be true. When the verb after *wish* is *BE,* use *were* with all subjects, singular and plural.

I **wish she were** here now. [She isn't here.]

Jake **wishes it were** warm today. [It isn't warm.]

We **wish we were** free to go. [We aren't free to go.]

In speaking, *was* is often used with singular subjects in sentences with *if, as if, as though,* and *wish.* For example, you will often hear statements like *If my brother was here now, we would have a family reunion* and *I wish she was here now.* In your writing, however, you should use the more formal *were* with all subjects.

practice 1

Rewrite each of the following sentences, adding the word in parentheses that correctly completes each sentence. Base your choice on formal written style.

1. The citizens' group asked that the bus company ___ two stops to its Main Street route. (add, adds)
2. Uncle Joe still talks to Ellie as if she ___ a child. (was, were)

3. If Terri's hair ___ a little darker, she would look just like her sister. (was, were)
4. Alex's counselor suggested that he ___ on several job interviews during summer vacation. (go, goes)
5. Don't you wish Danny ___ able to go with us tomorrow? (was, were)
6. Marty always behaved as though everything ___ all right. (was, were)
7. If Dad ___ a regular passenger, he would get to know the other commuters on the morning train. (was, were)
8. The tenants demanded that the landlord ___ the stairway immediately. (repair, repairs)
9. I feel as if I ___ walking on a cloud. (was, were)
10. Gene recommended that Nan ___ forward. (play, plays)
11. Dave acted as though he ___ not disappointed at the news. (was, were)
12. I wish that I ___ in grade twelve and graduating this June. (was, were)
13. The policeman recommended that every bicycle ___ reflecting lights. (have, has)
14. Diane tried to look as if she ___ having a good time. (was, were)
15. The judge insisted that the witness ___ the question. (answer, answers)
16. If Grandad ___ here now, we could go fishing together. (was, were)
17. Mom suggested that Wayne ___ to cook. (learn, learns)
18. Louis wishes the game ___ being played on Saturday instead of Friday. (was, were)
19. If Alexander ___ my dog, I'd enter him in a show this year. (was, were)
20. Connie suggested that each member ___ a dollar to the petty-cash fund. (contribute, contributes)
21. I wish I ___ as good a tennis player as Rosie is. (was, were)
22. We recommended that the ticket price ___ unchanged. (remain, remains)
23. I wonder what she would do if she ___ in my position. (was, were)
24. My married sister isn't living near us any more, but I wish she ___. (was, were)
25. The speaker continued as though he ___ not aware of the hecklers in the crowd. (was, were)

application

Write sentences of your own by completing each of the following expressions. Choose the verb form in parentheses for formal written style.

1. All the family would be together if Kate (was, were) ___.
2. Do you ever wish that he (was, were) ___?
3. Why does Mr. Kato act as if I (was, were) ___ ?
4. Her advisor recommended that Rita (take, takes) ___.
5. What would you do if she (was, were) ___?
6. Dr. Russo insisted that Dad (remain, remains) ___.
7. My brother still treats me as though I (was, were) ___.
8. Ms. Dravos suggested that Kevin (sit, sits) ___.
9. We all wish that Joe (was, were) ___.
10. Sometimes I feel as if I (was, were) ___.

27 Raise/Rise, Set/Sit, Lay/Lie

Is it *raise* or *rise? Set* or *sit? Lay* or *lie?* These three pairs of common English verbs seem to be persistent troublemakers. Because the two verbs in each pair have forms that look and sound somewhat alike and because their meanings are related, people frequently get them confused. However, each verb in each pair has its own meaning and use. It is not interchangeable with the other verb in the pair. Be sure to use the verb that carries the exact meaning you want to express.

Raise* and *Rise *Raise* means "to cause something to go up." It always is followed by an object and can be used in the passive.

Two scouts **raised** the flag carefully. [object—*flag*]

The flag **was raised** carefully by two scouts. [passive]

Rise means "to get up" or "to go up." It never is followed by an object, nor can it be used in the passive.

Please **rise** when the guest speaker enters. [no object]

Notice the difference in the forms these two verbs take:

raise	raising	raised	(have) raised
rise	rising	rose	(have) risen

Set* and *Sit *Set* means "to put or place something." It always is followed by an object and can be used in the passive.

Beth **set** the plants in the sun. [object—*plants*]

The plants **were set** in the sun. [passive]

Sit means "to take a sitting position" or "to occupy a place." It does not take an object.

The house **sits** on a small hill. [no object]

Notice the difference in the forms these two verbs take:

set	setting	set	(have) set
sit	sitting	sat	(have) sat

Certain expressions, established through long usage, are exceptions to these general rules—for example, *a hen sets, cement sets, the sun sets, a table sits ten,* and *a person sits a horse.*

Lay* and *Lie *Lay* means "to put or place something." It always is followed by an object and can be used in the passive.

We always **lay** the mail on that table. [object—*mail*]

The mail **is** always **laid** on that table. [passive]

Lie means "to recline" or "to be in a position or location." It does not take an object.

Grandmother always **lies** down after lunch for a nap. [no object]

Notice the difference in the forms these two verbs take:

lay	laying	laid	(have) laid
lie	lying	lay	(have) lain

You can see that the word *lay* is both the present tense form of the verb *lay* and the past tense form of the verb *lie.* Be careful. This shared form may lead you to use *laying* when you really mean *lying.*

Remember that *raise, set,* and *lay* are followed by an object and can be used in the passive; *rise, sit,* and *lie* are *not* followed by an object and have no passive forms.

practice 1

Rewrite each sentence, adding the correct form of *raise* or *rise* from the words in parentheses.

1. My aunt (raises, rises) turkeys as well as chickens.
2. The water level is (raising, rising) rapidly because of the heavy rains.
3. The booster's club has (raised, risen) nearly all the money it needs for new band uniforms.
4. In 1884 the Métis on the prairies (raised, rose) in rebellion.
5. Attendance has (raised, risen) since the boys started serving refreshments at club meetings.
6. Our morale (raised, rose) after we heard the good news.
7. Several hands were (raised, risen) when the speaker agreed to answer questions.
8. Two zoologists are (raising, rising) a baby chimpanzee in their home.
9. The sun (raises, rises) at 6:45 a.m. tomorrow.
10. Julian (raised, rose) the question of student parking facilities again.

practice 2

Rewrite each sentence, adding the correct form of *set* or *sit* from the words in parentheses.

1. Someone has (set, sat) a wet glass on the tabletop.
2. Mrs. Anderson invited us to (set, sit) down and rest.
3. The cabin (sets, sits) on a bluff overlooking the lake.
4. This plant should be (set, sat) in a strong light.
5. Last night Maria and I just (set, sat) and watched television.
6. We were (setting, sitting) the chairs outdoors when it started to rain.
7. Just (set, sit) the packages over there, please.
8. Everyone was (setting, sitting) in the wrong room and wondering where the teacher was.
9. Where should this trunk be (set, sat)?
10. After George had (set, sat) for three hours on the bus, he was tired and stiff.

practice 3

Rewrite each sentence, adding the correct form of *lay* or *lie* from the words in parentheses.

1. If you have a headache, why don't you (lay, lie) down for a while?
2. Mr. Garfield is (laying, lying) the carpeting in the hall today.
3. The cornerstone of the new library will be (laid, lain) next week.
4. Millie was (laying, lying) in a hammock, reading *Sports Illustrated.*
5. (Lay, Lie) the parcels on the counter in the kitchen.
6. The sailboat (laid, lay) at anchor near the pier.
7. The negotiator (laid, lay) his cards on the table.
8. The kitten has (laid, lain) asleep all afternoon.
9. New railroad track is being (laid, lain) along the old roadbed.
10. The bike path (lays, lies) along the course of the river.

application

Rewrite the following paragraph, adding the appropriate form of *raise* and *rise, set* and *sit,* or *lay* and *lie* from the words in parentheses.

> (Raising, Rising) a puppy is not easy. My brother and I have found that we have to protect anything chewable, like a shoe (laying, lying) on the floor. Also, we cannot (set, sit) plates of food in a spot the puppy might reach. Mom and Dad discovered another hazard when they were trying to (lay, lie) new tiles on the kitchen floor. (Laying, Lying) tiles is usually not difficult, but it is almost impossible with an inquisitive puppy underfoot. (Raising, Rising) costs of veterinary fees and dog food make it expensive to keep a puppy. But the effort will seem justified when we can all (set, sit) back and enjoy a healthy, well-trained dog as a member of the family.

28 Leave and Let

The verbs *leave* and *let* are frequently confused. Most often, *leave* is used improperly to mean "permit" or "allow," a meaning which actually belongs to the verb *let.*

*__Leave__ him go.

Let him go. [Allow him to go.]

*They **are leaving** us study together.

They **are letting** us study together. [They are allowing us to study together.]

In your writing, be sure you keep these two verbs and their meanings distinct. Don't use *leave* when you mean *let.*

Use *leave* when you mean "depart" or "cause something to remain."

They usually **leave** by eight.
He **left** school before graduating. } [depart]

We **are leaving** the rest for you.
I **have left** my umbrella at home.
Please **leave** the window open.
Luckily, the burn didn't **leave** a scar. } [cause to remain]

Use *let* when you mean "permit" or "allow."

Why don't you **let** them go to the movies?

We **are letting** the children stay up to see their grandparents tonight.

I **let** the cat out an hour ago.

They've **let** that mistake happen too often.

Don't hang on to that rope! **Let** go!

Notice the difference in the forms these two verbs take:

leave	leaving	left	(have) left
let	letting	let	(have) let

You can use either verb correctly with the word *alone,* as in these expressions:

Leave her alone. **Let** her alone.

*The asterisk (*) identifies incorrect sentences.

Here the two expressions have the same meaning—"Don't bother her." In all other uses, however, you should keep the meanings of the two verbs distinct.

practice 1

Rewrite each of the following sentences, replacing each blank with a form of *leave* or *let.* Choose the word that carries the meaning of the definition in parentheses.

1. Dad is ___ for Calgary on the six o'clock bus. (depart)
2. He is ___ me drive the car while he's gone. (allow)
3. Last night Alfonso ___ a note for his sister about the basketball game. (cause to remain)
4. Last night Alfonso ___ his sister know about the basketball game. (allow)
5. Since Karen is home, just ___ the back door unlocked. (cause to remain)
6. Will you ___ me out at the next intersection, please? (allow)
7. The police aren't ___ anyone through the barrier until the fire is under control. (allow)
8. I think you should ___ her finish the course if she wants to. (allow)
9. Olivia is hoping the doctor will ___ her go home tomorrow. (allow)
10. The team members had to ___ their baggage behind for the next bus. (cause to remain)

practice 2

Rewrite each of the following sentences, adding the right form of *leave* or *let* in parentheses.

1. Hang on! Don't ___ go! (leave, let)
2. Just ___ the books on the table. (leave, let)
3. Why don't you ___ them try to fix it themselves? (leave, let)
4. We have only three days ___ before vacation. (left, let)
5. Don't ___ the cat through the gate! (leave, let)

6. What time are you ___? (leaving, letting)
7. Why are you ___ the children stay up so late? (leaving, letting)
8. Please don't ___ your bike in the driveway. (leave, let)
9. They've ___ that problem get out of hand. (left, let)
10. Just ___ a message if I'm not home. (leave, let)
11. Will you ___ me help? (leave, let)
12. Don't ___ it bother you. (leave, let)
13. He ___ his watch to be repaired. (left, let)
14. I wish you'd ___ me explain. (leave, let)
15. Just ___ her tell the story herself. (leave, let)
16. We ___ her decorate the table. (left, let)
17. We ___ the table for her to decorate. (left, let)
18. Just ___ it to me. (leave, let)
19. You shouldn't ___ her get away with that! (leave, let)
20. Don't tease your brother; ___ him go! (leave, let)

← application →

Write ten sentences of your own, using the verbs indicated. Be careful not to confuse *let* and *leave.*

1. have let
2. leave
3. are letting
4. left
5. had let
6. is leaving
7. let
8. had left
9. is letting
10. will leave

29 Possessives with *–ing* Verb Forms

An *–ing* verb form like *swimming* or *laughing* may function as a noun. These verbal nouns, called gerunds, can appear in any noun position.

> **Jogging** can be hard work. [subject of sentence]
>
> Not everyone enjoys **cooking.** [direct object]
>
> Jamie was punished for **lying.** [object of preposition]

Even when these *–ing* forms function as nouns, they keep

some of the characteristics of verbs. For example, *–ing* forms may take a subject. When they do, the subject appears before the *–ing* form.

> **Her jogging** keeps Marilyn slim.
>
> We all enjoyed **Paul's cooking.**
>
> Everyone was surprised at **Jamie's lying.**

Notice that *Her, Paul's,* and *Jamie's* are in the possessive. In your writing, you should normally use possessives like these when the *–ing* form and its subject function as the subject for the entire sentence. This is especially true with personal pronouns and people's names that come at the very beginning of the sentence.

> **Our** finding an open gas station at that hour was pure luck. [not *Us finding*]
>
> **His** understanding the math problem right away was a surprise to us all. [not *Him understanding*]
>
> **Their** pretending not to care about the outcome didn't fool anyone. [not *Them pretending*]
>
> **Joan's** swimming in the relay improved the team's score considerably. [not *Joan swimming*]

However, in these situations, you would not normally use a possessive noun or pronoun before the *–ing* form:

1. The subject of the *–ing* form is a plural noun.

> In those days, Uncle Ted never approved of **women** voting. [not *women's*]

2. A phrase comes between the *–ing* form and its subject.

> We were surprised at the **owner** of a new car parking in such a dangerous place. [not *owner's*]

3. The noun subject of the *–ing* form does not refer to something living.

> Everyone talked about the **river** rising close to flood level. [not *river's*]
>
> They hoped that the **truth** appearing in the newspaper would stop the rumors. [not *truth's*]

4. The subject of the *–ing* form needs to be strongly emphasized.

> Can you imagine ***me*** getting the highest score on the history final? [not *my*]

practice 1

Rewrite each of the following sentences, adding the word in parentheses that correctly completes each sentence.

1. ___ threatening to make us do the cooking stopped our complaints about food. (Pat, Pat's)
2. Fortunately, no one asked for my opinion of ___ acting. (him, his)
3. ___ running the 100-yard dash in 9.1 seconds made Harry Jerome the world's fastest human in 1966. (Him, His)
4. We didn't understand the ___ of the mall objecting to the zoning regulations. (builder, builder's)
5. A sure sign of the beginning of the ragweed season is ___ sneezing. (Aunt Martha, Aunt Martha's)
6. The mayor feared that the ___ striking at night would add to the danger. (hurricane, hurricane's)
7. ___ calling the vet so promptly saved the collie's life. (Sharon, Sharon's)
8. The ___ crying brought worried members of the family from all rooms of the apartment. (baby, baby's)
9. Everyone was distressed by the ___ burning to the ground. (building, building's)
10. We appreciate ___ defending our viewpoint. (you, your)
11. In many Canadian homes, the idea of ___ helping with the housekeeping is no longer a strange one. (men, men's)
12. ___ pretending to understand is not going to help either of us. (Me, My)
13. We really didn't mind ___ interrupting. (you, your)
14. Most people today accept the idea of ___ participating in competitive sports. (women, women's)
15. The accident was caused by the ___ of the station wagon failing to see the red light. (driver, driver's)
16. ___ practicing the drums doesn't seem to bother the neighbors. (Jill, Jill's)
17. Mom was upset at ___ missing the bus again. (me, my)
18. The ___ keeping her kittens in the coat closet considerably changed our family habits. (cat, cat's)
19. ___ forgetting his lunch meant that he had to borrow money from me. (Tom, Tom's)
20. We had to go to a Laundromat because of the ___ breaking down. (washing machine, washing machine's)

← application →

Make a complete sentence of each of the following expressions by adding the possessive of a noun or a pronoun and an *-ing* verb form.

1. No one talked about ___
2. She didn't mind ___
3. I don't understand ___
4. He didn't explain ___
5. We were upset by ___
6. ___ pleased us.
7. ___ surprised everyone.
8. ___ was disappointing.
9. ___ seemed unimportant.
10. ___ didn't bother anyone.

30 Negative Words

In addition to *not* and its contraction *-n't,* English has many negative words, such as *no, none, never, nothing, no one, nowhere.* Also, prefixes like *in-, un-, dis-, non-* turn positive words into negative ones.

Sometimes using two negative words in a single sentence is entirely correct and appropriate.

> He **wasn't unhappy** about having to move.
>
> That expression is **not incorrect.**
>
> She has **never** been **unwelcome.**

However, in most of your sentences, be careful to use no more than one negative word in each clause.

> I **don't** want any pie. [one clause, one negative]
>
> I **don't** want any pie, and Bev **doesn't** either. [two clauses, two negatives]

Double Negatives In informal speech, more than one negative in a clause is sometimes used for special emphasis. But in writing, always avoid double negatives like this one:

> *We **don't** need **no** help.

Because English has so many negative words, there is often more than one way to write the same negative sentence. Notice that the following double negatives can be rewritten correctly in two different ways:

*The asterisk (*) identifies incorrect sentences.

*They **didn't** see **no one.** *We **don't** have **nothing** to do.
They **didn't** see **anyone.** We **don't** have **anything** to do.
They saw **no one.** We have **nothing** to do.

Notice also that the word *any* or a word containing *any,* such as *anything, anyone, anywhere,* is often used with a negative in a sentence.

Can't Hardly The words *hardly, scarcely,* and *barely* are considered negatives because they mean "almost none" or "almost not." If you use another negative in a clause with *hardly, scarcely,* and *barely,* you will be using a "concealed" double negative.

*I **can't hardly** understand you.
I **can hardly** understand you.

*He **couldn't barely** keep his eyes open.
He **could barely** keep his eyes open.

*There **wasn't scarcely** any food in the house.
There **was scarcely** any food in the house.

Ain't Be careful to avoid using *ain't* in your writing in place of *am not, is not, are not, has not, have not,* and their contractions. Remember that the contraction *aren't* may be used in negative questions with *I.*

*I **ain't** interested in going to that party.
I**'m not** interested in going to that party.
Aren't I going with you? I'm going too, **aren't I?**

*Julie **ain't** my cousin.
Julie **isn't** my cousin.
Julie**'s not** my cousin.

*He **ain't** left yet.
He **hasn't** left yet.

*They **ain't** here now.
They **aren't** here now.
They**'re not** here now.

*We **ain't** seen it.
We **haven't** seen it.

practice 1

Rewrite each of the following sentences, adding the word in parentheses that makes a correct negative sentence.

1. There isn't ___ easy way to explain what happened in the last quarter of the game. (no, any)
2. Toby worked all afternoon but couldn't find ___ wrong with the motor. (anything, nothing)
3. Once Norm starts talking about his motorcycle, there isn't ___ that can stop him. (anyone, no one)
4. Sally said she didn't want ___ special favors from us. (any, no)
5. We looked in the freezer for chocolate-chip ice cream but didn't find ___. (none, any)
6. Unless you get there by six, you won't find ___ to park. (anywhere, no where)
7. Can't you give me ___ reason for their strange behavior? (no, any)
8. The committee members don't want ___ to know about the awards before the presentation. (anyone, no one)
9. That cactus can live in a desert; it won't need ___ water for weeks. (any, no)
10. Only tourists go to Alcatraz Island now; there haven't been ___ prisoners since 1963. (no, any)

practice 2

Each of the following negative sentences can be written a second correct way. For example, *They didn't hear anyone* can also be written *They heard no one.* Changing only the words in heavy type, rewrite each sentence to form another correct negative sentence.

1. Wayne **hasn't ever** been to a track meet before.
2. I **didn't see anyone** who needed a ride to the stadium.
3. By eight-thirty there **wasn't anything** left to eat.
4. I **haven't** made **any** progress with my guitar lesson.
5. We talked for an hour but **didn't get anywhere.**
6. Callie looked for directions but **didn't find any.**
7. I **don't** want **any** more advice from them.
8. Tom **couldn't** find **anyone** to explain the road map.
9. **Isn't** there **anyplace** to hide the presents for Dad?
10. Aunt Gina **didn't see anything** she wanted to buy at the garage sale.

practice 3

Rewrite each of the following sentences, adding the negative word in parentheses and making any other changes necessary to avoid a double negative.

1. There isn't any milk left. (hardly)
2. Mike couldn't get to his seat before the kickoff. (barely)
3. I can't hear you when the stereo's on. (scarcely)
4. Al couldn't understand Annette when she spoke French. (hardly)
5. There isn't room for six of us in your VW. (barely)
6. There wasn't anyone there when Maria got to the clubhouse. (hardly)
7. She wasn't polite to me after I apologized. (barely)
8. I can't believe that I climbed that cliff. (hardly)
9. There wasn't enough kindling to start a fire. (scarcely)
10. The children can't wait for the puppet show to begin. (hardly)

practice 4

Rewrite each of the following sentences, removing *ain't* and substituting *am not, is not, are not, has not,* or *have not*—or their contraction.

1. I'm going with Mrs. Bashir, ain't I?
2. Ain't Mr. Kucera explained the homework yet?
3. I ain't going to lunch right now.
4. You ain't surprised at Jeff's attitude, are you?
5. Why ain't Charlene planning to compete in the 60-yard dash tomorrow?
6. Ain't you seen *Why Shoot the Teacher?* yet?
7. Pat ain't found the directions for making the kite yet.
8. Those are the headphones we're supposed to use, ain't they?
9. Those sportswriters ain't written enough about professional soccer.
10. Ain't the straight steal used more often than the delayed steal in softball?

← application →

Rewrite the following paragraph, choosing the right word from the words in parentheses.

> Some days I can't seem to do (anything, nothing) right. This morning I got up late and didn't have (any, no) time for breakfast. On the way to school it started to rain, and I didn't have (a, no) raincoat. There (was, wasn't) barely time for me to wipe off my wet books and to run to class. When I got there, I couldn't find (any, no) paper. Then I realized I didn't have (any, no) lunch. I couldn't find it (anywhere, nowhere). I had to buy my lunch, and I didn't see (any, none) of my friends to eat with. I'm glad there (aren't, ain't) too many days like today. And tomorrow I (am not, ain't) going to get up late.

31 Regular Adverbs

Most adverbs are formed by adding the ending *-ly* to an adjective. For example, the adjective *rapid* becomes the adverb *rapid**ly***, and the adjective *clear* becomes the adverb *clear**ly***. When you write a sentence that requires an adverb following the verb, don't use an adjective by mistake. Avoid writing a sentence like this:

*Dave works **efficient.** [needs an *-ly* adverb]

If the modifier following the verb describes the subject, then you must use its adjective form. However, if the modifier following the verb describes the action of the verb, you must use its adverb form.

Dave seems **efficient.** [adjective modifies noun *Dave*]

Dave works **efficiently.** [adverb modifies verb *works*]

A few adverbs like *loud* and *slow* may be used with or without the *-ly* ending. In formal usage, the *-ly* ending is usually preferred.

*The asterisk (*) identifies incorrect sentences.

Ned sang **loud.** Ned sang **loudly.**

Go **slow!** Go **slowly!**

Some adverbs like *fast* and *hard* have no *-ly* form.

They moved **fast.** You play too **hard.**

When you want verbs such as *taste, look, sound, smell, feel, remain, appear* to describe what something is like, use an adjective after the verb to modify the subject.

This line looks **crooked.**

The fresh air smells **good.**

The soup tastes **spicy.**

That music sounds **strange.**

This material feels **rough.**

However, when you want to use these same verbs to name an action, use an adverb after the verb to modify the verb.

Pat looked **suspiciously** at the strange box.

Janie smelled the newly baked bread **eagerly.**

The baby tasted the cereal **slowly.**

We jumped as the foghorn sounded **suddenly.**

Elva felt the hot iron **cautiously.**

To be sure you have used the right form after one of these verbs, try substituting *is, are, was,* or *were* for the verb in the sentence. If the sentence still makes the same kind of sense, an adjective after the verb is the right form. For example, test these two sentences by substituting *were* for *looked:*

The hikers looked **weary.** [adjective]
The hikers looked **wearily** up the trail. [adverb]

Since *The hikers were weary* makes sense, you know you need the adjective form *weary* in the first sentence. And since **The hikers were wearily up the trail* doesn't make sense, you know the adverb *wearily* is the right form for the second sentence.

practice 1

Rewrite each of the following sentences, choosing the word in parentheses that correctly completes the sentence.

1. Holly listened __ to her sister. (patient, patiently)
2. Holly was __ whenever she listened to her sister. (patient, patiently)
3. Be sure to answer __ if anyone asks you about the score. (polite, politely)
4. Sometimes Mr. Gerhart seems __ to strangers. (sarcastic, sarcastically)
5. The stage crew worked __ from 8:00 to 10:30. (steady, steadily)
6. Don't be __ when you take your driver's test. (reckless, recklessly)
7. If you drive __, you won't get your license. (reckless, recklessly)
8. The children played __ until dark. (noisy, noisily)
9. Watch __ while I turn on the motor. (careful, carefully)
10. Our dog always waits __ for us to take him for a walk. (quiet, quietly)

practice 2

Rewrite the following sentences, choosing the adjective or adverb form in parentheses that completes each sentence correctly.

1. Your idea sounds __. (sensible, sensibly)
2. We looked __ for the car keys. (hasty, hastily)
3. The fire alarm sounded __ and startled everyone. (sudden, suddenly)
4. They looked __ about the news. (angry, angrily)
5. Jake tasted the lemonade __. (cautious, cautiously)

6. The lemonade tasted ___. (sweet, sweetly)
7. The wood felt ___ before we sanded it. (rough, roughly)
8. We smelled the burning leaves ___. (happy, happily)
9. Joey felt ___ for the light switch in the dark. (hopeful, hopefully)
10. The hot rolls smell ___. (delicious, deliciously)

application

Use each of the following expressions correctly in a sentence of your own.

1. looked sadly
2. looked sad
3. stayed calmly
4. stayed calm
5. remained patiently
6. remained patient
7. appeared eagerly
8. appeared eager
9. felt cautiously
10. felt cautious

32 Good and Well, Bad and Badly

Choosing between the modifiers *good* and *well* or *bad* and *badly* is sometimes a problem when they come after the verb in a sentence. Is it *She sings* ***good*** or *She sings* ***well?***

Remember that in formal written style *good* is always used as a noun modifier, or adjective. Use *good* after such verbs as *BE, seem, feel, taste, smell, look, sound* when you want to describe a person or thing in a sentence.

Dinner smells **good.** [adjective]

The band sounds **good.** [adjective]

Those soccer players are very **good.** [adjective]

I feel **good** after a hot shower. [adjective]

When you want to describe an action in a sentence, use *well* as a verb modifier, or adverb.

Both Mother and Dad cook **well.** [adverb]

The band performed **well.** [adverb]

They played soccer **well.** [adverb]

You should also use *well* as an adjective when you are referring to someone's health or appearance.

Dan had the flu, but he is **well** again now. [adjective]

He says he feels **well** now. [adjective]

Rita looks **well** in that shade of green. [adjective]

Expressions like *feeling good* and *looking good* are often used in speaking to describe a general state of well-being. However, you should use the more formal *feeling well* and *looking well* if you really mean "not ill."

Use *bad* in the same way you use *good*—as a noun modifier after such verbs as *BE, seem, feel, taste, smell, look, sound.*

This milk tastes **bad.** [adjective]

The polluted river smells **bad.** [adjective]

Jack's first driver's test was **bad.** [adjective]

Use *badly* in the same way you use the adverb *well*—as a verb modifier that describes an action.

Dotty was nervous, and she sang **badly.** [adverb]

Ann plays softball **badly** but still loves it. [adverb]

Jack drove **badly** when he took his first driver's test. [adverb]

The expression *to feel badly,* meaning to be upset or unhappy, is widely used in speech. However, in your writing, remember to use the adjective form *bad* after *feel,* just as you would use other adjectives in the expressions *we feel sick, we feel upset, we feel unhappy.*

We feel **bad** about the damage done by the heavy rainstorm.

practice 1

Rewrite each sentence, choosing the word in parentheses that correctly completes each sentence in formal written style.

1. Marcy certainly looks ___ now that she has recovered from her cold. (good, well)
2. Don't you agree that the team played ___ yesterday? (bad, badly)
3. If you think this recording sounds ___, you can order one for yourself. (good, well)
4. Andy always looks ___ in a coat and tie. (good, well)
5. This pen writes just as ___ as the other one. (bad, badly)
6. If you speak French ___, you will certainly enjoy your trip to Quebec. (good, well)
7. Jeff can't see ___ without his glasses. (good, well)
8. Wayne's bike is as ___ now as it was when it was new. (good, well)
9. That movie wasn't as ___ as I thought it would be. (bad, badly)
10. Did you feel ___ when you heard the news? (bad, badly)
11. When Aunt Edie bakes bread, the whole house smells ___. (good, well)
12. The director scheduled another rehearsal today after the band performed so ___ last night. (bad, badly)
13. The patient responded ___ to the treatment of complete rest. (good, well)
14. The weather was so ___ that Mr. Neumann canceled the hike. (bad, badly)
15. Unfortunately, the motor doesn't run ___ when it rains. (good, well)

← application →

Use each of the following expressions in a sentence of your own.

1. feels well
2. are good
3. tastes bad
4. looks well
5. drove badly
6. feels bad
7. are well
8. played badly
9. runs well
10. smelled good

33 Adjective and Adverb Comparison

Almost all adjectives and adverbs have three different forms to show increasing degrees of the quality they name, as in *clear, clearer, clearest* and *awkwardly, more awkwardly, most awkwardly.* The positive form—*clear, awkwardly*—states the quality. The comparative form—*clearer, more awkwardly*—states an increasing degree of the quality. The superlative form—*clearest, most awkwardly*—states the highest degree of the quality. Because there are several ways to form these three degrees of comparison, choosing the proper form can sometimes be a problem.

Words of One Syllable Add the endings *-er* and *-est* to regular adjectives and adverbs to form their comparative and superlative degrees.

	Positive	*Comparative*	*Superlative*
Adjectives	tall	taller	tallest
	green	greener	greenest
Adverbs	soon	sooner	soonest
	fast	faster	fastest

Words of More Than Two Syllables Add the words *more* and *most* before regular adjectives and adverbs to form their comparative and superlative degrees. Also use *more* and *most* with all adverbs ending in *-ly.*

	Positive	*Comparative*	*Superlative*
Adjectives	ambitious	more ambitious	most ambitious
	unusual	more unusual	most unusual
Adverbs	unusually	more unusually	most unusually
	roughly	more roughly	most roughly

Words of Two Syllables Most two-syllable words use *more* and *most,* as in *eager, more eager, most eager.* But some, especially those ending in *-y,* may use either form of comparison, as in *foggy, foggier* or *more foggy, foggiest* or *most foggy.* If you are not sure which form to use with a two-syllable word, use *more* or *most.*

> Anne seemed **angrier** about the referee's decision than the other team members did.
>
> Anne seemed **more angry** about the referee's decision than the other team members did.

Irregular Forms Some frequently used adjectives and adverbs have irregular forms. That is, they do not form the comparative and superlative by adding either *-er* and *-est* or *more* and *most.*

Positive	*Comparative*	*Superlative*
good well	better	best
bad badly	worse	worst
much many	more	most
little	less	least

Notice that the adjective *good* and the adverb *well* share the same comparative and superlative forms—*better* and *best. Bad* and *badly* share *worse* and *worst,* and *much* and *many* share *more* and *most.*

Use of the Comparative The positive forms *much* and *many* may combine with comparative forms, as in *many more headaches, much more trouble, much faster, much more quickly.* But the comparative form *more* does not combine with another comparative, as in **more faster.* Be careful not to use double comparisons like **more better, *worser, *more heavier.*

Use the comparative form for comparing two items.

> Of the two problems on page 20, the second is the **more** difficult one. [not *most*]

Be sure also that the two things you compare can logically be compared with one another. For example, the following sentence is incorrect because you can't compare a test with biology, a subject of study:

> *This history **test** is harder than **biology.**

*The asterisk (*) identifies incorrect sentences.

The two things being compared must be made alike.

This history **test** is harder than a biology **test.**

When you compare a person or thing to the rest of the group it belongs to, use the words *any other* to set off the person or thing from its group.

*Sally is taller than **any** girl in her class.

Sally is taller than **any other** girl in her class.

Use of the Superlative The superlative form *most* does not combine with another superlative, as in **most fastest.* Be careful not to use a double superlative like **most best* or **most beautifulest.*

In speaking, the superlative form is sometimes used for the comparison of two things, as in *Both children are sick, but Janie is sickest.* In writing, however, use the superlative form for comparing three or more items.

Of the three problems on page 20, the second is the **most** difficult one. [not *more*]

practice 1

Rewrite each of the following sentences. Add the comparative or superlative form of the word in parentheses to complete each sentence correctly.

1. Although Leroy is ___ than his brother, they are both the same height. (young)
2. A giant Douglas Fir tree in British Columbia, more than 90 metres high, is said to be the world's ___ living thing. (large)
3. Of the four candidates, we thought that Stan had spoken the ___. (convincingly)
4. Which of those four record albums did you like ___? (well)
5. Golden Delicious apples usually ripen a few weeks ___ than Jonathan apples. (late)
6. The city of Quebec, founded in 1608, is the ___ city in Canada and the only walled city in North America. (old)
7. "But the second part of the story is even ___ than the first," said Naomi. (peculiar)
8. In 1970 high tides and a tropical cyclone hit

Bangladesh, creating the ___ natural disaster of the twentieth century. (bad)

9. Russell drove much ___ after he passed the driver's test. (calmly)
10. "That is the ___ sunset I've ever seen," said Laura. (spectacular)

practice 2

Rewrite the following sentences, choosing the right word or words in parentheses to complete each sentence correctly.

1. This chili is much ___ than the one we had last week. (hotter, more hotter)
2. Don't you think that test was harder than ___? (any we've taken this year, any other we've taken this year)
3. The ___ way to get there is by the tollway. (most quickest, quickest)
4. Edie's behavior is far more unpredictable than ___. (Jan's is, Jan)
5. Sid seemed ___ than I had ever seen him before. (more nervous, nervouser)
6. I liked all the food, but the chocolate cake was the ___. (bestest, best)
7. We are convinced that our spaniel Rex is smarter than ___. (any other dog on the block, any dog on the block)
8. "That was the ___ moment I've ever spent," confessed Lynn. (more awkward, most awkward)
9. Which river is ___, the Fraser or the Mackenzie? (longest, longer)
10. Danny said his cold was much ___ yesterday. (worser, worse)
11. "I'd rather go to Prince Edward Island than ___," declared Marusia. (any province, any other province)
12. Dave said that he had had the ___ trouble with the final chapter of the book. (least, leastest)
13. This recipe for cream-cheese cake is much easier than ___. (the apple pie, the one for apple pie)
14. The last day of classes always seems the ___. (slowest, most slowest)
15. Jeff's hair is longer than ___. (Sarah, Sarah's)

← application →

Write ten sentences of your own, using the comparative form of the words in numbers 1–5 and the superlative form of the words in numbers 6–10.

1. strange
2. good
3. dependable
4. difficult
5. smoothly
6. soon
7. bad
8. comfortable
9. quickly
10. rough

Mastering the Mechanics

34 Capitalization

The rules for capitalization are fairly well fixed, but if you're writing for a special purpose (the yearbook, a science paper), special rules may apply. Whenever you're uncertain, check the dictionary!

1. Capitalize the first word of every sentence, even if it is part of a name that is usually not capitalized.

Da Vinci was a great inventor.

Leonardo da Vinci lived in Florence.

2. Capitalize the first word in a direct quotation.

Mary asked, "How much was the radio?"

But if the quotation has been interrupted, do not capitalize the first word of the second part.

"How in heaven's name," she shouted, "do you expect me to get this theme in on time?"

If a quotation is not a complete sentence, begin it with a small letter.

What does the expression "icing the puck" mean?

3. Capitalize the days of the week and months of the year, but not the seasons.

Thursday	yesterday
October	autumn
Good Friday	birthday
Valentine's Day	holiday

4. Capitalize proper nouns and proper adjectives, which are the names of specific persons, places, or things.

Edmonton Eskimos	German recipe
Sominex	Maltese cross
Laura Secord	Okanagan apples
Torontonians	Victorian furniture
Coca-Cola	the Boer War
Montreal, Quebec	Dutch bulbs

However, some proper nouns and proper adjectives that once were capitalized no longer begin with capital letters.

bologna	manila folder
bowie knife	petri dish
french fries	plaster of paris
india ink	roman numeral

Originally these items were named after people or places. Through widespread use, these words have become common nouns and adjectives, and their capital letters have been replaced by small letters. The same process has taken place with some trademarks. *Thermos* used to be a particular brand of glass-lined bottles. Now *thermos* can be used by any company manufacturing glass-lined containers. However, *Band-Aid, Kleenex,* and *Xerox* are still trademarks, even though in everyday speech they are often used in place of *bandage, tissue,* and *to copy.* When you write, be sure to use capital letters for all registered trademarks.

5. Capitalize the regions of the country, but do not capitalize directions or points on the compass.

Sheila has lived in the West all her life.

The Graysons flew west to visit their relatives.

6. Capitalize titles with names, and also capitalize the titles when they stand in place of names.

General Montgomery   Senator Lawson
Reverend John Dawes

"Can you come in here, Mom?" "Welcome home, Professor."

Her mom and dad are both Chinese Canadians.

7. Capitalize the first word and all important words in the titles of books, poems, paintings, and the like.

My Darling, My Hamburger	*The Wind in the Willows*
"Mona Lisa"	"Ode on a Grecian Urn"
"Meet Me in St. Louis"	"Hot Buttered Soul"

8. Words like *school, hotel, church, street* are not capitalized unless they are part of a proper name.

the new hotel downtown	Four Seasons Hotel
the third street	804 West Third Street
high school	Lord Byng High School
their neighborhood church	St. Peter's Church

9. Don't capitalize *a, an,* and *the* unless they are part of a proper name.

the St. Louis *Post-Dispatch* *The Winnipeg Free Press*

practice 1

Rewrite the following phrases, adding capital letters where necessary. Some phrases do not require capitals.

1. fine french wines
2. speaks chinese
3. brazilian coffee
4. another shoe
5. the north
6. drive north
7. hamilton team
8. english bone china
9. premier peter lougheed
10. the new chairman
11. liberals
12. uncle joel
13. my youngest uncle
14. simplicity pattern
15. east on first avenue
16. the crimean war
17. old war hero
18. doctor nan fowler
19. king richard III
20. the ailing king

practice 2

Rewrite each sentence, adding the necessary capital letters. Forty-two capitals are needed.

1. when you get to watson road, turn northwest.
2. david and maureen took the twelfth street bus to dee's.
3. mrs. poe sent both her kids to camp runinmud for the whole summer.
4. the moscow circus is in toronto for the second week in may.
5. the film *why shoot the teacher?* though set in saskatchewan was made in alberta.
6. "what kind of climate does halifax have?" asked phyllis.
7. "how should I know," responded bill, "when I've never been east of the rockies."
8. two english families fought the war of the roses to determine who would be the next king.
9. "don't spend it all at the fair, andy," pop said.
10. larry said to luke, "pick a girl friend who'll let you tinker with your chevy in peace on saturdays."

← application →

Rewrite these paragraphs, changing small letters to capital letters wherever necessary.

> when brian found out great canadian oil sands wanted him to go to a meeting in edmonton, he asked me to come, too. we left fort mcmurray on a thursday morning, driving on the now completely paved highway. near wandering river we had a flat, but we inflated the tire with a can of pump-it-up and made it to a gas station. brian wasn't satisfied with his spare, so he bought a used michelin tire to replace the one that blew.
>
> as we rolled into edmonton on highway 28, now 97th street, we passed an atlas van lines moving van that had flipped over. furniture was scattered all over the intersection of 97th street and 137th avenue. city police were already at the scene, waving all traffic on. but that wasn't the last accident of the day. in the evening, while brian was attending a meeting, I watched an excellent production of othello at the citadel theatre. in the last act as othello was killing desdemona, the wall of their bedroom slowly fell down on top of them.

35 The Comma in Compound Sentences

You can express two ideas in two separate sentences.

Marilyn ran for office.
She won easily.

Or you can show that two ideas are closely related by joining them to form one compound sentence.

Marilyn ran for office**, and** she won easily.

Notice that two things are necessary to join ideas in a compound sentence—a comma and a connecting word, or *conjunction,* like these:

and but for nor or yet so

The conjunction connects the two ideas and shows how they are related. The comma before the conjunction provides a signal to the reader that one idea has ended and another has begun. Since the comma really acts as a separator, it can't be used *by itself* to form a compound sentence.

*Marilyn ran for office, she won easily.

In fact, if a compound sentence is very short, you may omit the comma before the conjunction.

It rained **so** we left.

However, before you leave the comma out, be sure that the sentence will not be confusing without it. Compare these sentences:

We enjoyed working for Mr. Tibe was a friendly boss.

We enjoyed working for Mr. Tibe was a friendly boss.

Which sentence is less confusing? Would you say that the comma is necessary here or optional?

practice 1

Rewrite these sentences, adding the necessary commas. Write *correct* after the number if no comma is needed.

1. Karen Magnussen won the World Figure Skating Championship and she later joined a professional ice show.
2. Joan strung the racket but it broke later.
3. Either you honor the contract or the company will be forced to take legal action.
4. Designs in Oriental rugs are symbolic and often the weavers of one tribe can't explain symbols from the rugs of another tribe.
5. Tammy wanted to frost her hair yet she was afraid the process might give her split ends.
6. We couldn't make much of a contribution nor did they expect a big one from us.
7. I was wrong and I'm sorry.
8. The dentist found one cavity but he decided to make another appointment for me to get it filled.

*The asterisk (*) identifies incorrect sentences.

9. Mom saw the mouse run in and out she went for traps.
10. Phil wanted to buy his girl friend a sewing machine for she was a good seamstress.

practice 2

Rewrite each pair of sentences as a compound sentence. Use commas and conjunctions correctly.

1. You could use compost to enrich this plot. You might choose chemical fertilizers for a faster effect.
2. Compost would improve the soil texture. Chemicals wouldn't help the texture at all.
3. Family Allowance cheques help thousands of Canadians feed their families. Without this program many would suffer.
4. Everybody with some income must pay for the scheme. Politicians complain that the program is a giveaway.
5. Kadonago has support from all segments of the community. He would win even without the youth vote.
6. He speaks well in public. His reputation for honesty is well known.
7. The government tests new medicines before they can be marketed here. The public often wants foreign drugs before they have been okayed.
8. Some people believe an extract from apricot pits can cure cancer. They import the product illegally.
9. Ann Landers writes an advice column. Her opinions on love and marriage are very popular.
10. She and her husband decided to get a divorce after thirty-six years together. She didn't try to hide this from her readers.

← application →

Write ten compound sentences of your own, showing the correct use of the comma before the conjunctions. Be sure to use every conjunction—*and, or, but, nor, for, so, yet*—at least once.

36 The Comma for Items in a Series

Use commas to separate items in a series, whether the items are words, phrases, or short clauses.

> Alice ordered folders**,** stationery**,** and stamps. [words]
>
> We hiked to the beach**,** around the lighthouse**,** and through the grove of trees. [phrases]
>
> He studied**,** he practiced**,** and he worried. [clauses]

You may omit the comma before *and* at the end of a series, but only if it is not needed to make the meaning clear. Newspapers and magazines often make it a practice to leave out all optional commas. But since using the comma before *and* is never wrong and not using it may cause problems, you'd be safer to put it in every time.

> Fresh air**,** the sight of green grass and good country food perked me up right away. [unclear]
>
> Fresh air**,** the sight of green grass**,** and good country food perked me up right away. [clear]

Some words are customarily used in pairs: *rights and privileges, bag and baggage, ham and eggs, pen and ink, bread and butter.* When these pairs occur with other items in a series, set them off with commas as just one item.

> Joey fixed soup**, bread and butter,** pudding**,** and milk.

But if you mean the items to be considered separately, put a comma between them.

> Reva bought soda**,** pickles**,** **bread,** and **butter**.

When all the items in a series are joined by *and* or *or,* no commas are used.

> Rex wagged his tail **and** yapped **and** shook for joy.

If a series of words seems to describe a single word that follows, you must check the final items in the series carefully for commas.

> The mayor's sister is a **tall, frail** woman.

Tall and *frail* both describe *woman,* so they are separated by a comma. But look at the word *dark* in this sentence:

> Ellie crocheted her mother a **dark** green sweater.

Dark tells what kind of *green,* doesn't it? Since *dark* modifies *green,* these two words should not be separated by a comma. Note that there is never a comma between the last modifier and the word it modifies.

There are two purely conventional uses for the comma in a series of items. Commas separate items in a date or an address and set them off from the rest of the sentence.

> The bombing of Hiroshima, August 6, 1945, brought an end to the war with Japan.
>
> The Hollys moved to 707 Wineva Avenue, Toronto, Ontario, last year.

The last comma in these sentences is required.

practice 1

Rewrite these sentences, supplying the necessary commas. If the sentence is correct as given, write *correct.*

1. The Dawsons the Gattis the Podjarskis and the Janssens play bridge every Thursday.
2. That laundry will sew on missing buttons repair zippers and do reweaving.
3. Lou left his money to his sons his house to Mrs. Kane and his stocks to Marie's trust fund.
4. Molly served ham and eggs rolls and tea at her brunch.
5. Mrs. Tolliver's portfolio included common stocks municipal bonds and sinking-fund debentures.
6. The decorator matched the wallpaper with the sofa the carpet and the dining room chair seats.
7. Felix Bev and Carl share the ride to school.
8. The October 1 1972 issue of *Saturday Night* was sold out.
9. Many movie stars live in Palm Springs California between their film assignments.
10. She modeled a crisp dove-gray taffeta skirt.
11. Mike and Sherry Nick and Donna and Jack and Leann are the other couples I've invited to the pool party.
12. The medical secretary makes appointments prepares bills and orders supplies.
13. The Channel 2 weatherman predicted sleet or snow or hail for the three-day forecast.
14. Who is the dark handsome man over by the fireplace?

15. The dark brown horse is being ridden by a nervous inexperienced jockey.

← application →

Write ten sentences of your own, using these series of items. Be sure each series is punctuated correctly.

1. white roses orange glads and green ferns for contrast
2. Mom Dad my two uncles and Puff
3. Toronto Quebec City and Ottawa
4. July 1 1867
5. sturdy construction attractive styling good safety features and a money-back guarantee
6. the library the post office and the zoo
7. awkward anxious and shy
8. brave and trustworthy and loyal
9. the July 1975 issue
10. farm in York Township Ontario

37 The Comma—Introducers/Interrupters

Commas set off words that are not part of the main structure of a sentence.

Introducers Use a comma to set off introductory elements from the rest of the sentence.

> **When the tide went out,** the men dug clams. [clause]
>
> **Running easily,** the fox outdistanced the dogs. [phrase]
>
> **After a long illness,** Mary returned to work. [phrase]
>
> **Luckily,** the dog was found before Kim had to start the painful series of rabies shots. [word]

If the introductory element is a short prepositional phrase like *After a long illness,* you may omit the comma. But using the comma is never wrong.

No comma is necessary if the normal word order of the sentence has not been disturbed.

> The men dug clams **when the tide went out.** [clause]
>
> Mary returned to work **after a long illness.** [phrase]

But clauses starting with *though* or *although* should always be set off with commas, whether they come at the beginning or the end of a sentence.

Although the cafe was open, they didn't have any tenderloin sandwiches.

We enjoy going pheasant hunting**, though we seldom have any luck.**

Interrupters Use commas to set off material that interrupts the general flow of a sentence.

Cory will be admitted**, I think,** after he gets a pass.

You know**, of course,** that his job is in jeopardy.

If interrupting material is in parentheses and a comma would naturally follow, place the comma right after the parentheses. In this example you need a comma after the introductory *when* clause:

When we left (at sunrise or a little later), the party was still in full swing.

In dialogue, the words identifying the speaker are also set off by commas from the material being directly quoted.

"I never told you**,**" **she cried,** "you could leave without finding me another ride."

But when the quoted material is a question or an exclamation, you'll need to use question marks or exclamation points instead of commas at the end of the quotation.

"Get out of that closet**!**" Don shouted at the dog.

"Did your subscription start yet**?**" asked Elayne.

Be sure that punctuation which is part of the quoted material is placed *inside* the quotation marks.

practice 1

Rewrite these sentences, setting off the introducing or interrupting elements with commas. Two sentences are correct as given. For these, write *correct* after the number.

1. When the map was finally located it led us to an empty, underground cavern.
2. Before victory was assured the candidate shook thousands of hands and kissed hundreds of babies.

3. Larry did finally get to the theater though he was too late for the previews or cartoons.
4. The company found so I'm told that their new product was not selling well in the west.
5. Considering how many families do not eat breakfast it's no wonder so many kids fall asleep in first period.
6. "They won by cheating" said Rajesh.
7. After the first half (the score was then 31–10) the Lions got discouraged.
8. Palestrina to my way of thinking is the greatest composer before Bach.
9. Because the decorating budget was limited I had to settle for cafe curtains instead of wooden shutters.
10. "Detective Holt lost the suspect in Union Station" said the sergeant "but I saw him later."
11. Marv remarked "I thought you were back hours ago."
12. The best course to follow in my opinion is to give Ms. Dell a raise and some additional responsibilities.
13. The cleanup began after the last group of businessmen had left the convention area.
14. As you may already know the bowling tournament had to be postponed until next Sunday.
15. Use liquid bleach on all whites except nylon and rayon.

← application →

Proofread these paragraphs. Then rewrite them, adding commas to set off introducers and interrupters.

> When you have to make a long-distance move get several estimates. (All companies do *not* charge the same rates though the government does regulate movers on other matters.) The estimate is not a bill of course but just an idea of your total cost for the move. The actual price you pay will depend on how much your shipment weighs. Naturally nobody can estimate that just by looking at your furniture.
>
> If you want to you can have the movers pack *everything.* But the most common practice is to pack some things yourself and leave some items for the pros to pack. When the movers do the packing they furnish boxes and stuffing materials, but you have to pay for

them. If you explain your wishes to the estimator all the appropriate packing charges will be put into the estimate.

To assure fair treatment for yourself try to be present for the weighing of your shipment. Assuming the mover you chose is reputable (and most are) you should get a fair shake.

38 The Comma – Nonessential Elements

Commas are used to set off nonessential sentence parts—words, phrases, or clauses—from the main structure of a sentence. Essential words, phrases, or clauses are *not* separated from the rest of a sentence.

How can you tell whether a sentence part is essential or not? If the word, phrase, or clause answers the question *which particular one?* about the person or thing it refers to, it is essential to the meaning of the sentence. It should not be set off by commas. For example, the following sentence contains an essential element:

> We need a flower **that blooms well in the shade.**

Without the clause in color, the sentence means that *any* flower will do. The clause is essential to identify *which particular kind* of flower is needed. In other words, the clause in color restricts *flower* to one specific kind. Because of this, such essential elements are often called *restrictive.*

If the word, phrase, or clause does not identify the word it refers to as one specific person or thing, it is *nonessential.* It should be set off from the rest of the sentence by commas. For example, the following sentence contains a nonessential element.

> The impatiens**, which blooms well in the shade,** is available in all colors of the red-to-white range.

In this sentence the flower's name, *impatiens,* answers the question *which particular flower?* The clause in color is not essential to identify the particular kind of flower. *All* impatiens bloom well in the shade. The clause in color provides additional information about *impatiens* but does not restrict or limit the meaning of that word in any way. In fact, the clause could be omitted without changing the basic meaning of the sentence. Such nonessential elements are often called *nonrestrictive.*

Here are some examples of essential and nonessential words and phrases. Pay special attention to where and to why commas are used.

My friend **LaVerne** is up for treasurer. [essential word]

Gaston **, my best friend,** is from Sudbury. [nonessential phrase]

The gift **wrapped in expensive red paper** is yours. [essential phrase]

The gift **, wrapped in expensive red paper,** took her last penny. [nonessential phrase]

Any team **in the tournament for the first time** is at a disadvantage. [essential phrase]

Our team **, in the tournament for the first time,** is at a disadvantage. [nonessential phrase]

Remember that all *nonessential* elements are set off by commas.

practice 1

Decide if each of the following sentences includes a nonessential element. If it does, rewrite the sentence, adding the necessary commas. If not, write *correct* after the number. There are four correct sentences.

1. Mr. Bixby our neighbor left his lawn mower outside.
2. My sister and her best friend Ellen Dukes bought their tickets together for the reserved-seat concert.
3. The cat in the third cage may have feline distemper.
4. Hodgkin's disease which is a form of cancer often strikes the lymph nodes first.
5. A mother with seven children is entitled to relax at the end of the day.
6. The White Rock Players a well-known theatrical group gave a special performance of their Christmas pantomime for crippled children.
7. A dentist who tells his patients to use dental floss regularly is doing them a favor.

8. *Moby-Dick* which some critics call the Great American Novel has been filmed more than once.
9. Churchill's mother who was born in America married into British aristocracy.
10. The pin heavily encrusted with rubies and seed pearls was complemented by the black velvet box it rested in.
11. My new Pacer which has a wiper on the hatchback window is a well-equipped car.
12. The latest figures with still only a few polls reporting show that the Liberals are leading.
13. Jackie Dawson playing bridge for the first time outscored all other contestants in the novice event.
14. The cottage next to ours at Lake Simcoe is up for sale.
15. The Minister of Finance who has control over your tax bill has just recommended a personal income tax increase.

practice 2

First write ten sentences of your own, using each of the following word groups as an essential element. Then write ten more sentences, using the same word groups as nonessential elements. Be sure to set off nonessential elements with commas.

1. finished by hand
2. who ran for the NDP
3. thinking of his girl
4. which don't need much space
5. whose mother is Jewish
6. taken in jest
7. laughing his head off
8. with jelly on her face
9. known for her honesty
10. which comes in blue

← application →

Proofread this paragraph carefully. Then rewrite it, setting off all *nonessential* elements with commas.

> One of Vancouver's tourist attractions in 1977 a baby beluga whale was born on July 13 1977 at the Vancouver Public Aquarium. The actual birth was captured on film that was shown on television sets around the world. While beluga whales have been born in captivity before the Vancouver newcomer is the first in North America to survive for more than a few hours.

Aquarium officials waited anxiously for a month to see whether the baby would indeed survive before announcing a "Name Our Baby" contest a contest that attracted over 10,000 entries. The winning name was Tuag an Eskimo word meaning "the only one." The contest winner Denis Foster said he picked the name out of a reference book. The story which should have ended happily ended in fact in tragedy. Tuag was found lying dead at the bottom of his pool by Klaus Michaelis the aquarium's chief trainer.

39 The Semicolon

The semicolon is a stronger separator than a comma but not so powerful as a period. A semicolon cannot end a sentence. But it can do something neither the comma nor the period can do. It can join two sentences together to form a compound sentence without the help of a connective word. Semicolons can also be used instead of commas to separate items in a series when one or more items have commas of their own.

Use a semicolon between two sentence structures not connected by *and, or, but, for, nor, yet,* or *so.*

> The automakers predicted a record-breaking year**;** Chrysler's sales figures went up 28% in January.

For the semicolon to be appropriate, the sentences it joins must be closely related in meaning. If the sentences are not related, separate them with a period.

If either sentence already contains a comma, use a semicolon to join the sentences, even when words like *and, but, or, so* are present.

> Lt. Vaughan**,** a Calgary native**,** scored 570 points**; but** one of the female recruits did just as well later.

When two sentences are linked by conjunctive adverbs like *however, therefore, in fact, for example, on the other hand,* you must use a semicolon *before* the connectives. (A comma usually follows these connectives.)

> She invested shrewdly**; however,** she lost everything when gold stocks collapsed.

> There are several good reasons**; for example,** you've never seen the city, and you don't know the company.

Note that the material after the semicolon *must* make a complete sentence by itself. If it doesn't, then use a comma instead of a semicolon.

> Clare has many good points, **for example,** poise, talent, ambition, and intelligence.

Another use of the semicolon is to separate items in a series if any of them contain commas.

> Joy knows Hindi, the official language of the North; Tamil, a Ceylonese tongue; and Sanskrit.

practice 1

Rewrite the following sentences, changing incorrect commas to semicolons. If a sentence is punctuated properly, write *correct* after the number. There are three correct sentences.

1. They moved often, living in Calgary, Alberta, Winnipeg, Manitoba, and London, Ontario, during the 1970s.
2. She fell out of the window, but I didn't push her.
3. Atlanta is growing rapidly, it offers a good business climate and a stable working population.
4. Billy, who is just three, ate his brother's candy, and this caused a terrible row when Henry got home.
5. Pencils don't contain lead anymore, they're made of graphite, which isn't poisonous.
6. Supply and demand govern prices, so when goods are in short supply and great demand, prices go up.
7. Neil's victory was no surprise, he's been practicing that dive for six months.
8. Jenny has a collection of antique bottles, let her explain the difference between pressed and cut glass.
9. The profit-earnings ratio is good, therefore I recommend that you buy the stock as soon as possible.
10. A new store would bring in more business, however, we just don't have enough capital to expand now.
11. Clean your work area, and then lock the door.
12. Dad replaced the screens just in time, the seventeen-year locusts invaded the very next night.
13. Karen scored well in the school figures, she lost only because Nancy's free-skating routine was so very good.

14. Grandpa bought a new tiller, some blood meal, and special enzyme tablets, which will break down the composted vegetable matter faster.
15. Nick keeps jumper cables in his car, so we don't really have anything to worry about.

application

The following paragraphs contain commas used incorrectly. Proofread and rewrite the paragraphs, replacing the incorrect commas with semicolons.

> The cosmetic industry is a powerhouse in the North American economy, it accounts for billions in sales every year. The cosmetic giants—Revlon, Max Factor, Clairol, Estee Lauder—give one another stiff competition. They need new products all the time, they must keep an "exclusive" image, they must finance expensive research.
>
> For example, to bring out a new moisturizer, the chemists must formulate emollients, which are skin softeners, humectants, which hold moisture to the skin surface, and binders to keep the cream smooth. The product must smell and look good, it must have a shelf life of eighteen months before spoilage, it must be attractively packaged by the advertising agents. If any one of these qualities is missing, the new moisturizer will fail in the marketplace. Ballpoint lipsticks, makeup-setting mists, one of which was nothing but water, and lash-extending mascaras are just three famous marketing flops in the industry.

40 The Apostrophe for Possession

Until about 300 years ago, the apostrophe was not used to show ownership. Possessives were simply written with the letters *s,* the same as most plurals. Today, though, you must use an apostrophe to indicate ownership.

To make a singular noun or an indefinite pronoun show ownership, add *'s*.

the boss's daughter	Kathleen's ring
Ms Travis's coat	the team's mascot
everyone's duty	somebody's purse

Remember that the possessive pronouns never have apostrophes, even though some end in *s*.

my / mine	your / yours	her / hers	his
our / ours	their / theirs	whose	its

Most plural nouns already end with the letters *s* or *es*. So to make them show ownership, add just an apostrophe.

the boys' bikes	five carpenters' toolboxes
the Burches' house	soldiers' rations

If a plural noun ends with a letter other than *s*, make it possessive by adding *'s*.

the men's room	children's toys

Certain expressions of time and money are always written with apostrophes, although these apostrophes do not indicate ownership in the usual sense.

a dollar's worth	two months' work
a week's vacation	three hours' relief

practice 1

Rewrite these phrases, making the nouns in heavy type possessive by adding *'s* or just an apostrophe.

1. **fox** tail
2. **teacher** pen
3. **women** room
4. **ladies** dresses
5. **Joann** typewriter
6. **plumbers** wrenches
7. **leaders** demands
8. **doctor** recommendations
9. **cat** pajamas
10. **winner** number
11. **driver** techniques
12. **car** accessories
13. **oxen** yokes
14. **nobody** business
15. **anybody** guess
16. **Louise** scarf
17. **Liberals** strategy
18. a **month** rent
19. **bird** nest
20. **Tess** predicament

application

Rewrite these sentences, adding possessive nouns that end with an apostrophe or *'s.*

1. I left __ tablet at home this morning.
2. Mother donated half __ allowance to the fund drive!
3. How can she get the job done on five __ notice?
4. The __ barn burned to the ground in thirty minutes.
5. __ sewing machine breaks the thread too often.
6. The __ room is a mess; they should clean it up today.
7. I left my __ son reluctantly at the nursery-school door; Sherry couldn't drop him off that day.
8. I put the __ appointment book beside the switchboard.
9. That __ clothes are always in the height of fashion.
10. Dad makes good pizza, but my __ pies are even better.
11. The __ victory was unexpected considering their record.
12. The cat clawed up __ social studies homework.
13. The __ limousine was a '73 Cadillac.
14. Though the __ race is expected to be close, Morris will probably win.
15. She found __ necktie in the laundry hamper.

41 Apostrophes – Contractions/Plurals

In addition to indicating possession, the apostrophe has two other common uses. It takes the place of missing letters in contractions. Also, the apostrophe is used to form the plurals of figures, letters, and symbols.

Contractions—shortened forms of common expressions—are acceptable both in spoken English and in informal written English. Use an apostrophe to take the place of letters left out when you write a contraction.

we're [we are]	he'll [he will]	it's [it is]
she's [she is]	she's [she has]	I've [I have]
can't [cannot]	won't [will not]	they're [they are]

Another use for the apostrophe is in forming special plurals. Most of the time you can make nouns plural by adding just the letter *s.* But to form the plurals of figures, letters, symbols, and words referred to as words, you should add *'s.*

The picture order included some 5 x 7**'s.**

Jane got B**'s** and C**'s** on all her math homework.

The +**'s** represent areas of rain on the forecast map.

You have too many *because***'s** in that sentence.

When you choose to write out numbers and symbols, no apostrophe is needed.

Willie can count by **fives** to one hundred; he's only six.

Lee got two **minuses** in reading.

practice 1

Rewrite these sentences, making contractions of the phrases enclosed in parentheses.

1. (She had) arrived in Kingston long before the storm.
2. The dress (I have) chosen is light brown wool.
3. Are you sure (he will) come to the study session?
4. Rehearsal starts at eight, but (they are) often late.
5. (Let us) try the new bypass south of Winnipeg.
6. (We have) been planning the party since October.
7. (She has) never been elected to office before.
8. That geranium (does not) need pinching to bloom well.
9. (They have) taken the math test already.
10. (He is) the new tenor in the senior choir.

practice 2

Rewrite these sentences, adding an apostrophe or *'s* where it is needed. Some sentences need two apostrophes.

1. With the order for three 8 x 10s the photographer included some wallet-size pictures at no extra cost.
2. Lilly's third-grade report card had seven +s and no −s.
3. Ive gotten As on all the important assignments so far.
4. When you add 100s and 1000s, be sure to line up all the digits carefully.
5. How many *s* are there in *Mississauga?*
6. The type slug is broken, so all the 8s are blurry.
7. The deejay gave away six new 45s as prizes.
8. The *s in this column mean that we have gone over the figures at least twice.

9. To find substitutes for those extra *think*s in paragraph one, look up some synonyms in the thesaurus.
10. Uncle Max stockpiled some 2 x 4s in anticipation of the wood shortage that didn't come until summer.

← application →

Proofread the following paragraph. Then rewrite it, supplying the necessary apostrophes.

> Ive found that spoken language, with its regional differences in pronunciation and local expressions, is often more fresh and interesting than standard written English. New Yorkers, for example, "graduate high school," while the rest of us "graduate *from* high school." Chicagoans make their *couldnt*s into one-syllable words. Montana natives use a contraction youll almost never hear elsewhere: *maynt* for *may not.* Many Americans drop final *r*s, but Bostonians are the only ones with enough nerve to drop *r*s in the middle of words. They "pahk" their "cahs." People from Brooklyn, on the other hand, add *r*s where they dont exist. They "berl" water, as Archie Bunker says; the rest of us *boil* it. What many Canadians do not realize is that we too have our regional differences in language. For an entertaining look at the subject, read Mark M. Orkin's *Canajan, Eh?*

42 Quotation Marks—Direct Quotations

Use quotation marks when you record someone's exact words.

> Paul said, "My bike has a flat tire." [direct quotation]

Don't use quotation marks if you are reporting in your own words what someone has said.

> Paul said that his bike had a flat tire. [indirect quotation]

Often in an indirect quotation, the word *that* is omitted.

> Paul said his bike had a flat tire. [indirect quotation]

In writing quotations, you usually need to identify the

speaker with words like *he said* or *Linda announced.* The example sentences that follow show you four ways to do this. Notice that these quotations begin with capital letters and are set off from the sentences in which they appear by commas. Notice also that when commas and periods come next to quotation marks, they always come *before* the quotation marks.

1. Linda announced, "Nobody should miss that movie."
2. "Nobody should miss that movie," Linda announced.
3. "Nobody," Linda announced, "should miss that movie."
4. "That movie is great," Linda announced. "Nobody should miss it."

In example 1, the words identifying the speaker come before the quotation; in example 2, they follow the quotation. In both cases, these words are set off from the quotations by commas. In example 3, the identifying words interrupt the quotation. They are set off by a pair of commas, and the quotation continues with a small letter. In example 4, the identifying words also interrupt, but they mark the end of a sentence. So they are followed by a period. The quotation continues as a new sentence and begins with a capital letter.

Quotation marks are also used to record dialogue—conversation between two or more speakers. Each time the speaker changes, you should begin a new paragraph.

"Does it end happily ever after?" asked Sue, sitting up suddenly. Her eyes were wide open now.

"You know," said her father.

"No, I don't," she protested.

"She was a princess, wasn't she?" he asked.

"Of course."

"Well, then," he replied, "it just has to end happily ever after, doesn't it?"

practice 1

The following sentences have been written without punctuation. If the sentence needs quotation marks, write *direct quotation* after the number of the sentence. If the sentence does not need quotation marks, write *indirect quotation.*

1. Jamie said that he would take the train
2. Jamie said I'll take the train
3. Linda shouted that she was leaving
4. The dentist said it wouldn't hurt a bit
5. Carol said I know you can do it
6. The driver told us we were on the wrong bus
7. The driver told us You are on the wrong bus
8. Bill promised I won't do it again
9. Bill promised he wouldn't do it again
10. Sally replied I practice every afternoon

practice 2

Rewrite each sentence, adding quotation marks where they are needed.

1. Toby said firmly, I don't want any dessert.
2. This story, remarked Sara, is better than the last one.
3. It's not as long, either, Max replied.
4. I don't think, Barbie remarked, that he should have said that.
5. Beth whispered, We're planning a surprise for Ronnie.
6. I'm afraid, Jill replied, that he already knows about it.
7. That's not bad, observed Dad. I couldn't have done better myself.
8. I'd rather have pizza than spaghetti, declared Kim.
9. That's the trouble, said Ali. We don't know the way.
10. Well, at last we're ready, said Mom. Let's go.

← application →

Proofread the following dialogue. Rewrite it, adding quotation marks where they are needed.

> Don't worry, said Mom calmly. There's no point in getting married unless you're really sure. Granny used to say that if you marry in haste, you'll repent at leisure.
>
> I smiled up at her. I thought to myself how Granny used to use these folksy sayings all the time. Meanwhile Mother went on ironing.

How long did you know Bruce before he proposed? she asked.

Oh, about seven months, I said.

Did you like his parents? Mom asked.

That's just it, I said. I felt like *they* didn't approve of *me.* His father didn't say two words all through dinner.

What kind of people are they? Mom asked.

They're pretty well off, I guess. When I went up to one of their paintings, his mother asked me if I knew it was an original. I told her I knew.

Mother winced. That's my daughter, she said, a monster of tact.

Anyhow, I went on, it was all downhill after that. When I told Bruce I didn't think we could make it, he actually looked a little relieved.

There was a short pause.

Well, you know that Granny used to say you don't just marry a man. You marry a family.

Yeah, Mom, I said, and I sure didn't want to marry his mother!

43 Quotation Marks & Other Punctuation

Direct quotations are always enclosed in quotation marks. But where do commas, periods, semicolons and other marks of punctuation belong in relation to quotation marks?

Commas and periods always go *inside* the closing quotation marks. However, semicolons and colons go *outside* the quotation marks because they are not considered part of the quoted material.

> She said, "My girl is too good for Mr. Carson"; then she swept out of the room, her head high.
>
> The critic said, "An understanding of Shakespeare is obviously not enough": apparently we have to love the Bard, as well.

If the quoted material is a question or an exclamation, its question mark or exclamation point goes inside the closing quotation mark.

> She responded by saying, "What if it happens to you?"

But if the whole sentence is a question or an exclamation, put the question mark or the exclamation point *outside* the closing quotation mark.

> Did Mr. Chaffee say, "Read Chapter Two"?
>
> I could scream when she starts with "If I were you"!

You may need to write a sentence where the quoted material and the whole statement *both* should end with question marks or exclamation points. In this case, use just one question mark or exclamation point, *inside* the closing quotation mark.

> Do you remember her asking, "Can anybody give me a lift to the train station?"

Occasionally a quotation may occur inside another quotation. When this happens, use *single* quotation marks around the *inside* quotation.

> "Why do I get yelled at," Harry asked, "when Dad says 'Shut up!' as often as he wants?"
>
> Ms. Kelly said, "For Tuesday, write a paraphrase of Keats's poem 'Ode on a Grecian Urn.' "

Besides enclosing dialogue, quotation marks have two conventional uses. They enclose the titles of short works—magazine articles, short poems, essays, stories; songs; and the names of artworks. (Titles of longer works—books, long poems, plays, movies—and the names of ships are italicized. In handwritten or typed manuscripts, this is done by underlining.)

> Did you know Van Gogh is the subject of Irving Stone's novel Lust for Life and Don McLean's song "Starry Starry Night"?

Quotation marks also enclose words being defined or referred to as words.

> In my English theme I used "irony," but the right word was really "satire."

practice 1

Rewrite these sentences, adding quotation marks wherever they are needed.

1. Cleo said, You did better than Ivy on the history test.
2. They caught the vandals who spray painted the steps last Saturday, Lorraine announced.
3. Milton's *Paradise Lost* is hundreds of pages long, but he also wrote shorter poems, like On His Blindness.
4. I'll drop that overdue book off when I go to the grocery, said Mom, if it would save you a trip.
5. Lawrence's short story The Rocking-Horse Winner is about a little boy who has the gift of prophecy.
6. If you don't get out of here, I'll call the police! shouted Granny at the burglar.
7. The dentist said, After the shot, you won't feel a thing; he was right, too.
8. How I hate it when he says I told you so!
9. Avoid using ain't: being in the dictionary doesn't make a word acceptable in writing.
10. When Bobby ran over to the stereo, Julia said, Don't make the needle jump!

practice 2

Rewrite the following sentences, turning them into direct quotations. You will have to add speaker-identification words, like *she said,* and quotation marks. Change the double quotation marks given into single quotation marks.

1. "Day Tripper" is my favorite Beatles song.
2. Can you sing the harmony on "Beautiful Dreamer"?
3. But the critics liked Kate Reid's performance in the movie *The Andromeda Strain.*
4. Do you really expect me to know all the details of Rembrandt's "Night Watch" after seeing it twice?
5. Get the job done, or you're finished with my company!
6. Don't use "anxious" when you really mean "eager."
7. Tomorrow we'll hear a recording of Earle Birney's poem "David."
8. Don't let Meeper on the sofa with those muddy paws!
9. I read about it in the *Harper's* article "Power to the People: How the Electric Companies Run Your Life."
10. Will can explain what's wrong with the expression "between you and I."

← application →

Write ten original sentences, supplying the information called for and the punctuation required for each sentence.

1. a sentence with the titles of two short poems
2. a sentence that uses and defines a technical word
3. a sentence referring to the title of an article and the magazine it comes from
4. a quotation containing the title of a movie
5. a question ending with a quotation
6. a quotation inside another quotation
7. a statement ending with a quotation that includes a question mark
8. a sentence with the title of a musical play or movie and the title of one of its songs
9. a sentence with a quotation followed immediately by a semicolon
10. a sentence ending with an exclamation point, including a quotation anywhere in the sentence

44 The Colon

A colon in a sentence is a signal to the reader that something important follows.

Use a colon to introduce a list when you don't use the words *for example* or *such as.*

> The cake called for unusual ingredients**:** mace, citron, and coffee.
>
> Our bill included the following**:** telephone charges, telegrams, and room service.

However, do *not* use a colon when the list immediately follows a verb or a preposition.

> Beth's favorite foods **are** peanuts, salami, cheese, and hot fudge sundaes. [no colon after **are**]
>
> Amy sent invitations **to** band members, players, coaches, and cheerleaders. [no colon after **to**]

Use a colon between two statements when the second statement explains or restates the first.

> The printer rejected the apprentice's work: all the copies were smeared with ink.
>
> Today has been terrible: my ring broke, my math homework got lost, and my English paper came back with a *C.*

Colons also have some purely conventional uses.

1. To introduce a long or formal quotation.

   In her review of the new film, Rona Barrett says: "*One Flew over the Cuckoo's Nest* is about mental institutions. By the end of the movie you will be thinking of the cuckoos in your own nest."

2. To separate hours from minutes.

   8:15 a.m. 2:12 E.S.T. the 6:20 bus

3. To follow the greeting in a *business* letter.

   Dear Sir: Gentlemen: Dear Ms. Adams:

practice 1

Rewrite each sentence, adding colons where necessary.

1. We caught the bus at 330.
2. Harry missed the 504 GO Train to Oshawa, so he waited until after 630 for his wife to arrive with the car.
3. We ignored the confusion a new restaurant on opening day is always in chaos.
4. At the *Reader's Digest* office, letters that start *Dear Carolyn Davis* get sent to the subscription department because Ms. Davis is the subscription computer!
5. Phil gave several good reasons for staying in Victoria the winter climate, his high salary, and his low mortgage rate.
6. Dad added four items to the list checkbook, lug wrench, picnic tablecloth, and tent pegs.
7. It takes a long time to become a doctor four years of college, four years of medical school, and one or two more years of internship.

8. You may quote from only these books *Compton's, World Book,* and the *Farmer's Almanac.*
9. Ruth Stout invented a foolproof system for weeding she mulches the garden so weeds can't grow at all!
10. This quotation shows author John Davies (1610) was a male chauvinist "Deeds are males, words females are."

← application →

Proofread the following paragraph. Then rewrite it, adding the five needed colons.

> The actors arrived on flight 7 at 615. At 620 troubles started for the troupe No cabs were available, baggage disappeared, and the hotel lost their reservations. Later the prop master and the stage manager argued about who was responsible for some missing properties. Finally a list was posted. It read, "Please find the following before tonight one pair blue tights, one tiara, one velvet cape, one shovel." When the tights finally showed up, they were much too small for the star. He had to give Hamlet's famous speech, "To be or not to be that is the question," while standing behind a table to keep his legs hidden from the audience.

45 Parentheses, Dashes, Brackets

Occasionally, you may want to insert into a sentence words that sharply interrupt its normal word order. In such cases you will need stronger separators than commas. There are three special marks of punctuation you can use to set off or enclose these abrupt interruptions. The punctuation you choose depends on how strong a separation you want to indicate.

Parentheses Parentheses indicate the greatest degree of separation between the enclosed word group and the rest of the sentence.

Cyclamen (called "poor man's orchid") makes a good houseplant for cool rooms.

If the word group in parentheses occurs within a sentence, don't start it with a capital letter or end it with a period, even if the word group is a complete sentence.

Jim's wool jacket (we bought it last week) keeps him warm in sub-zero weather.

But if the word group is a question or an exclamation, place a question mark or exclamation point inside the parentheses.

That play (it *was* a comedy, wasn't it?) left me cold.

If the material in parentheses is a complete sentence standing by itself, begin it with a capital letter and end it with a period inside the closing parenthesis.

We left Saskatoon early. (It was just in time, too.) The hailstorm hit soon after ten o'clock.

Dashes Use dashes to indicate greater separation than commas, but less than parentheses. Dashes mark an abrupt change of thought or structure in a sentence.

Muhammad Ali—star of talk shows, amateur poet, heavyweight champ—has just written his autobiography.

Use a single dash to emphasize an added comment.

Ken could win a gold medal—provided he stays at 126 pounds.

Use a dash to do the work of a colon when a colon seems too formal.

Jerry Baker says all these are needed to grow a healthy lawn—several types of seed, frequent "shampoos," and regular additions of fertilizer.

If the interrupting word group is a statement, no period is needed inside the dashes. But if it is a question or an exclamation, you must add a question mark or exclamation point before the second dash.

The great Paul Newman—can you believe he's *fifty*?—shuns publicity and lives a quiet life.

Brackets Brackets have only two common uses. In quoted material they show that a clarification or explanation has been added. For example, the bracketed words in this sentence provide a clear reference for a pronoun.

> Discussing the theories of the Canadian communications specialist, Mr. Martin said: "When he [Marshall McLuhan] first said 'the medium is the message' he may not have realized how controversial the statement would be."

Brackets are also used to set off interrupting elements that occur in material already enclosed by parentheses.

> Our losses at Monte Carlo (we gambled away five hundred mostly at blackjack]) were made up in Nice.

Parentheses and brackets tend to give your writing a scholarly, formal tone. Dashes suggest surprise or emotion and give your writing a more casual tone. Don't overuse parentheses, dashes, and brackets; and don't use them as careless replacements for commas.

practice 1

Rewrite these sentences, adding needed parentheses.

1. Humus decayed plant matter improves garden soil.
2. Jack "Happy" Evans owns a restaurant in Kitchener.
3. Maxwell House "good to the last drop" and Boban "Nabob spelled backwards" are popular brands of coffee.
4. Aunt Lil's present it was custom made came from the best store in Montreal.
5. The carolers appreciated their host's wassail a hot, spicy punch and eggnog.
6. Hot-rod accessories glass-pack mufflers, headers, etc. took Hal's entire allowance.
7. Agatha Christie's detective wasn't Albert Finney great in the role? always solves the case creatively.
8. Rita and Yuu Chin opened a law office together. They're U.B.C. grads. Their practice is mainly divorce cases.
9. Berg polled 17% of the vote. He ran as an Independent. Such a good showing is rare for the splinter parties.

10. Two restaurant critics bought a dinner in Paris it cost $4000! that included nine different wines.

practice 2

Rewrite these sentences, supplying dashes and brackets. One sentence also needs a question mark.

1. Chuck made the most elusive golf shot a hole in one on the back nine at Capilano.
2. Bordeaux mixture and Paris green these are not exotic French compounds but common garden preparations.
3. All these make Ms. Dodge a fine boss to work for good humor, patience, and tolerance.
4. Wines are rated on bouquet how they smell as well as on flavor.
5. The dressmaker details on that blouse pin tucks, bound buttonholes, welt seams make it a bargain at $17.
6. Beverly Johnson she was *Vogue*'s first black cover girl makes well over $250,000 a year.
7. The Jordans adopted a sixth child isn't that great in addition to their five natural kids.
8. P. T. Barnum said, "There's one a sucker born every minute."
9. Several musicals have been about ethnic minorities *Purlie, Milk and Honey, West Side Story* but *Oliver!* is the only one about poverty.
10. Nureyev and Baryshnikov both Russian, both defectors are the world's greatest male dancers.

← application →

Write ten sentences of your own, using the following expressions to illustrate the correct use of parentheses, dashes, and brackets.

1. —I admit I'm addicted to them—
2. (Boy, was that a relief!)
3. He [the Prime Minister]

4. (one of the all-time greats)
5. (who could forget those eyes?)
6. —at an incredible $7.95—
7. (six of the fourteen survivors [three were women])
8. —all these make a good teacher.
9. (never one of my favorites)
10. —she actually went there!—

46 End Punctuation

Periods, question marks, and exclamation points usually signal the end of a sentence. Such end punctuation also tells the reader what kind of sentence he has just finished reading—a statement, a question, or an emphatic expression.

If you want to indicate a statement of fact or opinion, end your sentence with a period.

> I think the school newspaper should come out only twice a month so we can stay within our budget**.**

If you want to indicate a direct question, use a question mark as end punctuation.

> Do you think we should put the paper out every week or only twice a month**?**

But an indirect question hidden in a statement should be ended with a period.

> The principal asked me **if we should change to a semimonthly publication schedule.**

Polite requests may be punctuated with periods, especially if the answer is not much in doubt.

> Would you close the door as you leave, Fred**.**
>
> May we expect payment by the fifteenth**.**

However, a question mark for these sentences would be perfectly proper, too.

Reserve the exclamation point for use in sentences that express very strong feelings. Here are some sentences in which the exclamation point is used correctly.

> Wow! What a fantastically lucky break that trade was for the Blue Bombers!
>
> How that dog's constant yapping irritates me!

A command could end with either a period or an exclamation point, depending on the force you want to convey.

> Don't let the screen door slam, Molly. [mild]
>
> Don't tell Mother on me! [more forceful]

In all cases the exclamation point indicates greater than normal emphasis. Use this mark sparingly, especially in formal writing, like most school assignments.

As a general rule, end punctuation goes *inside* any quotation marks that may be present.

> Joe said, "I wish I could get these verbs memorized before French class."
>
> Maria wondered, "Can I finish this and still make it to wood shop on time?"

But question marks or exclamation points that refer to the whole sentence, not just the quotation, go *outside* the quotation marks.

> Did you finish reading "The Raven"?

practice 1

Rewrite these sentences, supplying the appropriate end punctuation.

1. Kolek won by the slimmest of margins
2. Can you believe that Kolek won
3. How fantastic that Kolek finally won
4. Get the plumb line out of the toolbox, Lenny
5. May I have New Year's Eve off if I work on Christmas
6. May we expect to hear from you by return mail
7. Your credit rating may be damaged if you let this bill go until next month
8. If you wait right there, I'll join you in a few minutes
9. Will you wait for me if I'm late
10. Get out of the doorway, for heaven's sake
11. Tim called angrily over his shoulder to the boys, "That'll be the day"

12. Kate asked softly, "Don't you want me to be your best friend any more"
13. Have you noticed the similarity between Tom Thomson's painting "Jack Pine" and A. J. Casson's "White Pine"
14. The girls wondered whether to plan the barbecue for a Friday or a Saturday
15. How incredible that he just said, "OK by me"

practice 2

Proofread the following paragraphs. Then rewrite them, supplying the needed periods, question marks, and exclamation points. Be sure to capitalize the first word of each new sentence.

When Mr. Denton came home from work, his wife set some TV dinners on the table and said, "Ben, we have to have a talk" the TV dinners should have been warning enough, but do you think Ben Denton paid any attention to them he did not he did, however, address his Salisbury steak and fries with interest

"Ben," Mrs. Denton began her husband dutifully looked up from his plate "I want to get a job" Ben's fork sank slowly down to the foil tray

"You've got a job," he said "You're my wife"

"That's not what I mean I want a job that pays money, a real job"

"Ina, what can you do—take in washing" Mr. Denton sounded hostile "I want you here when I get home," he went on "I want Tory to have a mother at home"

"Oh, Ben, Tory's a senior next year he'll be away at college what makes you think I shouldn't have a life of my own, outside the house?" she asked

← application →

Write at least ten sentences that continue the dialogue between Mr. and Mrs. Denton. Be sure to use a variety of sentence types—questions and exclamations as well as statements. Every time the speaker changes, start a new paragraph.

47 Numbers

Most of the time numbers in sentences should be spelled out. However, use figures for dates, hours (with a.m. or p.m.), addresses, page numbers, exact sums of money, and measurements expressed with common abbreviations.

July **1, 1867** [but *The **first** of July was overcast.*]

8:00 a.m. or **8** a.m. [but **eight** o'clock]

a **$19.95** blouse [but *Can you change a **fifty** for me?*]

722 E. **38th** Street Highway **401** on page **54**

−7° C **3″ × 5″** cards a **50′** drop

Use words to write round numbers or indefinite numbers. Also, spell out the ordinal numbers, which express order or position.

a **hundred** shares of stock a **million** chances to win

fourth Prime Minister **fourth** student in line

You should spell out any number that can be written in one or two words. (The numbers from twenty-one to ninety-nine are always hyphenated.)

twenty-six years old **eight thousand** immigrants

But use figures for numbers that would require more than two written words.

174 acres a **13.9** percent increase

8,200 subscribers a **106**-year-old Indian

Spelled fractions are hyphenated if they are used as adjectives, but not hyphenated when they are used as nouns.

Bac Tho won by a **three-fourths** majority. [adjective]

Three fourths of the class voted in the election. [noun]

You should never start a sentence with figures. Either spell out the numbers, or recast the sentence so the figures come in the middle or at the end.

*1967 saw the Canadian Centennial celebrations and a World's Fair in Montreal.

Nineteen sixty-seven saw the Canadian Centennial celebrations and a World's Fair in Montreal.

The Canadian Centennial celebrations of **1967** were accompanied by a World's Fair in Montreal.

If one item in a series requires the use of numerals, list every item with figures. All the numbers in the following example could be spelled out in two words or less, except 124. So all the numbers are expressed as figures.

The girl guide troop ordered **50** boxes of Savannahs, **73** of Chocolate Chips, and **124** of Thin Mints.

practice 1

Rewrite the following sentences, changing any numbers that are expressed incorrectly. Five sentences are correct as given. For these, write *correct.*

1. We adopted 6 stray cats over the years.
2. Nell ordered 24 crocus bulbs, 36 narcissus bulbs, and 150 Red Emperor tulip bulbs from the nursery.
3. The Martinellis moved to 3354 Drexel Avenue last year.
4. Vicky cancelled her 5 o'clock appointment and asked Roxanne to squeeze her in around 10:30 a.m. on Saturday.
5. You can enter Vancouver on 99A after going through New Westminster or take 99 through the Massey Tunnel, which is faster.
6. 0° on the Celsius scale is +32° F.
7. Elvis probably got 1,000 marriage proposals a year.
8. Our 1st trip to Ontario was uneventful, but on our 2nd vacation there we won $10,000 in the provincial lottery.
9. You can elect to take your full government pension at age 65 or a reduced benefit at age 60.
10. The skirt pattern called for a 6 centimetre hem.
11. One of the most attractive selling features of the house is its three quarter-acre lot.
12. The doctor got my blood pressure down to one hundred sixty-eight over one hundred ten with medication.
13. CP Air flight 604 leaves for Toronto and Montreal at 9:40 a.m. daily except Sundays.

*The asterisk (*) identifies incorrect sentences.

14. 1978 found Canada struggling with inflation and a high rate of unemployment.
15. My birthday falls on the 23rd of December, and nobody pays much attention to it because of Christmas.

application

Write fifteen sentences of your own, using each of the following number phrases. Be sure to change to spelled-out numbers if figures are not correct.

1. 15½″ neck and 33″ sleeves
2. 80 days
3. a forged $50 bill
4. 1976 [at the beginning of a sentence]
5. 1929 [in the middle of a sentence]
6. a 3-toed sloth
7. 27 new tomato hybrids
8. at 6:15 a.m.
9. June thirteenth, 1965
10. 1,000s of insects
11. our 25th anniversary
12. 2 cups of flour and 1 tsp baking powder
13. 6 pike, 13 sunfish, and 34 small perch
14. 3/5 of the men over 50
15. a 5 metre pole vault

Improving Spelling Skills

48 Spelling by Syllables

Suppose you want to write a long word like *competition* or *environment,* but you're not sure how to spell it. Instead of trying to spell the whole word at once, try an easier way—syllable by syllable. Say the word and count the vowel sounds. Usually, there is one vowel sound for each syllable. Dividing a word into syllables gives you small word sections that are easy to work with, like **com pe ti tion** and **en vi ron ment.** Write the word, repeating the syllables aloud as you spell one syllable at a time.

Here are three common patterns that are helpful for dividing words into syllables:

Consonant / Consonant	*Vowel / Vowel*	*Vowel / Consonant*
won / der	ne / on	ho / tel
struc / ture	ri / ot	ti / ger

Dividing a word into syllables does more than make the word easier to spell. It also helps you avoid two frequent causes of incorrect spelling.

The first cause is adding a sound or syllable that isn't really part of the word. For example, the word *athlete* is often misspelled because it is pronounced as if it had three syllables—**athelete*—instead of two. If you divide it into syllables—*áth / lete*—and pronounce it carefully, you can hear that it has only two syllables.

A second cause of misspelling is omitting a sound or syllable that really is part of the word. For example, the word *history* is often misspelled because it is pronounced as if it had only two syllables—**histry*—instead of three. If you divide it into syllables—*hís / to / ry*—and pronounce it carefully, you can hear that it has three syllables.

practice 1

Say the words in the following list and decide how many syllables each word has. Then write each word, drawing a

*The asterisk (*) indicates incorrect spelling.

line between the syllables. Use a dictionary to check your answers.

1. basis	5. final	9. license	13. rodeo
2. domestic	6. fluid	10. pliers	14. society
3. entertain	7. forty	11. pursue	15. stadium
4. except	8. identity	12. quiet	16. vitamin

practice 2

The following words are sometimes pronounced with an extra syllable. Write the words, grouping them into four lists according to whether they have one, two, three, or four syllables. Use a dictionary to check your answers.

athletics	disastrous	evidently	lightning
barbarous	drowned	hindrance	remembrance
burglar	entrance	hungry	umbrella

practice 3

One word in each of the following pairs is misspelled because of a missing syllable. Say each pair aloud. Then write the word whose spelling accounts for all the syllables.

1. accidentally|accidently
2. choclate|chocolate
3. dangerous|dangrous
4. defnite|definite
5. diffrent|different
6. favorite|favrite
7. genral|general
8. incidently|incidentally
9. mathmatics|mathematics
10. memory|memry
11. miniture|miniature
12. probly|probably
13. separate|seprate
14. sevral|several

application

Each of the following sentences has two words that contain a dash. If these words need extra letters, write the words, adding the extra letters. If these words are correct, write them without the dashes.

1. Sev—ral bicycles are blocking the ent—rance to the driveway.
2. Although it was a disast—rous flood, no animals were drown—ed.
3. My sister's fav—rite subject is math—matics.
4. Incident—ly, Mr. Stavros has a new mini—ture poodle.
5. The scouts were so hung—ry after hiking that each one bought a choc—late bar.
6. By mistake, Kim left her umb—rella and brought home a diff—rent one.
7. Evident—ly our high school has one of the most complete ath—letics programs in the province.
8. At last, def—nite plans have been made to correct that dang—rous curve in the Trans-Canada Highway.
9. Prob—ly the light—ning frightened the children.
10. Grandad's mem—ry is usually good, but he has no rememb—rance of the accident.

49 Words with *ie* and *ei*

When the letters *e* and *i* occur together in a word, it's hard to know what order to put them in—*ie* or *ei.* Most words having this letter combination are spelled with *ie.* A smaller number of words have *ei.*

How do you decide which combination to use for a particular word? The old rule expressed as a rhyme still helps, even though there are some words that are exceptions to it. (The symbol /ā/ stands for the "long a" sound in *able, pane, neighbor, and say.*)

Use *i* before e, except after c,
Or when sounded like /ā/,
As in *neighbor* and *weigh.*

Here are some examples of words that follow the rule, as well as three exceptions:

i before **e**	**ei** *after* **c**	**ei** *for /ā/*	*Exceptions*
believe	ceiling	beige	efficient
chief	conceit	freight	height
friend	receive	vein	their

practice 1

The following words are spelled correctly. Rewrite the words, arranging them into four lists. Head the lists i before e, ei after c, ei for /ā/, and Exceptions.

ancient	mischief	seize
brief	neighborhood	series
deceive	receipt	species
foreign	reign	weight
leisure	relief	yield

practice 2

Rewrite the following words, completing each by adding *ie* or *ei* according to the rule. There are no exceptions to the rule in this list. Check your answers with a dictionary.

1. ach—ve	6. front—r	11. r—ns
2. conc—ve	7. gr—ve	12. rel—ve
3. dec—t	8. hyg—ne	13. rev—w
4. d—sel	9. perc—ve	14. unv—l
5. f—rce	10. p—ce	15. windsh—ld

application

Rewrite each sentence, replacing the word or words in heavy type with the word in the list that has a similar meaning.

brief	freight	hygiene	relieve	unveil
deceive	frontier	leisure	review	yield

1. The railroad cars were at the dock to receive the **goods for transportation.**
2. A contract agreement was announced after each side agreed to **give up possession of** some of its demands.
3. Fortunately, we had only a **short** wait in the bitter cold outside the theater.
4. With a shorter workweek, many people will have more **free** time for hobbies and adult education.
5. The company's lawyer claimed that their advertising was not intended to **mislead** the public.

6. To **bring about the removal of** the threat of famine, the government sent rice to the flooded area.
7. The study of **the establishment and maintenance of health** is a required part of our science program.
8. The mayor will **remove the covering from** the statue at the Remembrance Day services next week.
9. Life on the western **edge of developed territory** held many hardships for the early pioneers.
10. Mrs. Shen urged us to **study again** the chapter on the 1930s before the exam.

50 Doubled Consonants Before Suffixes

If you want to write the past tense forms of *stop* and *stoop,* you need to add the suffix *-ed* to each. Why do you write *stopped* with two *p*'s but *stooped* with only one? When do you double the final consonant of a word before adding a suffix?

You double the final consonant when *all three* of the following conditions are present:

1. The suffix you are adding begins with a vowel—for example: *-ing, -ed, -able, -er, -est, -y.*

forget + ing = forgetting

2. The word you are adding a suffix to ends in a single consonant preceded by a single vowel.

begin + ing = beginning

3. The word has only one syllable, or the final syllable is stressed. Stress is indicated by this mark (´).

pin + ed = pinned pre/fer + ed = preferred

Remember, *all three* of these conditions must be met if you are to double the consonant. In other words, *don't* double the consonant if *any one* of these conditions applies:

1. The suffix you are adding begins with a consonant.

forget + ful = forgetful

2. The word ends in two consonants or in two vowels and a consonant.

return + ing = returning proceed + ing = proceeding

3. The stress doesn't fall on the last syllable, or the stress moves from the last syllable when the suffix *-ence* is added.

án/swer + ed = answered con/fér + ence = cónference

practice 1

Combine each word with the suffix indicated. Write each new word, using the three-part rule to decide whether you should double the final consonant before adding the suffix.

1. plan + ed
2. remít + ance
3. occúr + ed
4. rob + ery
5. compél + ing
6. regrét + able
7. sun + y
8. admít + ance
9. refér + al
10. propél + ing
11. big + er
12. jog + ing
13. bag + age
14. red + est
15. drag + ed

practice 2

Combine each word with the suffix indicated. First, decide if all three conditions necessary for doubling consonants are present. Then, write the new word correctly.

1. meet + ing
2. repel + ed
3. exist + ence
4. omit + ing
5. commit + ment
6. refer + ed
7. drug + ist
8. benefit + ed
9. expel + ed
10. regret + ful
11. fun + y
12. differ + ent
13. commit + ed
14. prefer + ence
15. transmit + al
16. rebel + ion
17. control + able
18. occur + ence
19. confer + ed
20. refer + ence

application

Proofread the following paragraphs carefully. Then rewrite the paragraphs, correcting any spelling mistakes you have found. There are eight misspellings of words with suffixes.

Flag football as an alternative to touch football has become increasingly popular with both men and women. In this variety of football, the equivalent of tackling has occured when a player removes the plastic streamers from the ballcarrier's belt.

Flag football is often prefered for intramural programs. It is frequently played by those who like to participate in sports but who are not totally commited to intense competition or to wining. In fact, players with enthusiasm but with differring degrees of ability are welcomed.

Players of flag football have sufferred fewer injuries from this variety of the game. They are also begining to enjoy a biger part in an active sports program.

51 Suffixes After Final *e*

When you add suffixes to words ending in *e* like *care* and *love,* why do you write *careful* and *lovely* with the *e* but *caring* and *lovable* without it? When do you drop the final *e* before adding a suffix? There are three general rules that apply to a large number of words ending in *e.*

1. If the suffix you are adding begins with a vowel, you drop the final *e.*

dance + ing = dancing
guide + ance = guidance
explore + er = explorer
scene + ic = scenic
sense + ible = sensible
criticize + ed = criticized
prepare + ation = preparation
ridicule + ous = ridiculous

2. If the suffix you are adding begins with a consonant, you keep the final *e.*

hope + less = hopeless
use + ful = useful
safe + ty = safety
extreme + ly = extremely
improve + ment = improvement
decisive + ness = decisiveness

3. If the word ends in *ce* or *ge,* you keep the final *e* before adding *-able* or *-ous.*

notice + able = noticeable
manage + able = manageable
courage + ous = courageous
outrage + ous = outrageous

Unfortunately, not all words that end in *e* follow these general rules. Here are some exceptions.

In spelling the following words, you keep the *e* before a suffix beginning with a vowel:

dye + ing = dyeing　　mile + age = mileage
singe + ing = singeing　　hoe + ing = hoeing

However, you drop the *e* before a suffix beginning with a consonant in spelling these words:

argue + ment = argument　　awe + ful = awful
whole + ly = wholly　　true + ly= truly
judge + ment = judgment　　nine + th = ninth
acknowledge + ment = acknowledgment

In spelling most words ending in *le,* you drop the *le* before adding *-ly.*

possible + ly = possibly　　probable + ly = probably
double + ly = doubly　　incredible + ly = incredibly

And in a few words ending in *ie,* you drop the *e* and change the *i* to *y* before adding *-ing.*

ie + ing = lying　　die + ing = dying　　tie + ing = tying

practice 1

Combine each word with the suffix indicated. Write each new word, following the three rules for adding suffixes to words with final *e, ce,* and *ge.* There are no exceptions in this list.

1. advise + ed
2. definite + ly
3. continue + ous
4. retire + ment
5. desire + able
6. chose + en
7. nine + ty
8. interfere + ence
9. service + able
10. advantage + ous
11. separate + ly
12. value + able
13. change + able
14. nerve + ous
15. arrange + ment
16. write + ing
17. expense + ive
18. knowledge + able
19. like + ly
20. decide + ing

practice 2

Combine each word with the suffix indicated. Write each new word, following the rules for adding suffixes to words with final *e, ce,* and *ge.* There are twelve exceptions in this list.

1. behave + ing
2. argue + ment
3. charge + able
4. pursue + ing
5. tie + ing
6. immediate + ly
7. awe + ful
8. probable + ly
9. make + ing
10. nine + th
11. persuade + ed
12. enforce + able
13. whole + ly
14. peace + able
15. judge + ment
16. use + ing
17. mile + age
18. lose + er
19. acquire + ed
20. lie + ing
21. possible + ly
22. true + ly
23. acknowledge + ment
24. appropriate + ly

← application →

For each of the numbered words or expressions, write the word in the list that has a similar meaning. Use a dictionary if you need help with any of the meanings.

chosen	guidance	nervous
continuous	improvement	ninth
courageous	interference	pursuing
dying	manageable	ridiculous
expensive	mileage	safety

1. increased excellence
2. uninterrupted
3. absurd, laughable
4. condition of being free from danger
5. facing death
6. ordinal number between eighth and tenth
7. selected
8. total distance in miles
9. easily excited or irritated
10. brave

11. having a high price
12. legal blocking of an opponent in football
13. following in order to overtake
14. capable of being handled with skill
15. process of leading or directing another

52 Suffixes After Final *y*

If you add the suffix *-ed* to words ending in *y* like *deny* and *obey,* why do you change the *y* to *i* to write *denied* but leave the *y* unchanged to write *obeyed?* When do you change a final *y* to *i* before adding a suffix? Two simple rules can help you add suffixes correctly to words like *deny* and *obey.*

1. When the final *y* is preceded by a consonant, as in *deny* or *worry,* you change the *y* to *i* before adding the suffix.

worry + ed = worried　　funny + er = funnier
ready + ness = readiness　　easy + ly = easily
try + es = tries　　baby + es = babies

However, if the suffix you are adding begins with *i,* like *-ing* or *-ish,* leave the *y* unchanged.

worry + ing = worrying　　forty + ish = fortyish

2. When the final *y* is preceded by a vowel, as in *obey* or *relay,* you leave the *y* unchanged.

relay + ed = relayed　　play + er = player
enjoy + able = enjoyable　　employ + ment = employment
convey + s = conveys　　decoy + s = decoys

There are a few exceptions to these rules. However, these exceptions normally cause little trouble.

The rules do not apply to the plurals of proper nouns or to nouns which add the possessive ending *'s.*

the **Kellys**　　the **baby's** blanket

In the following words, even though a vowel precedes the *y,* change the *y* to *i:*

day + ly = daily　　gay + ly = gaily

And in these words, the final *y* remains, even though a consonant precedes it:

shy + ly = shy**ly**	shy + ness = shy**ness**
sly + ly = sly**ly**	sly + ness = sly**ness**
dry + ly = dry**ly**	dry + ness = dry**ness**

Depending on its use in a sentence, the word *dry* has two different spellings when the suffix *-er* is added.

drier weather [adjective] a clothes **dryer** [noun]

practice 1

Combine each word with the suffix indicated. Write each new word, following the rules for adding suffixes to words with final *y*. There are no exceptions in this list.

1. early + est
2. hurry + ing
3. library + an
4. ready + ness
5. sorry + er
6. employ + able
7. study + ing
8. destroy + ed
9. empty + ness
10. company + es
11. annoy + ance
12. lucky + er
13. alley + s
14. ally + s
15. study + ed
16. convey + ance
17. rely + ance
18. busy + ness
19. try + ed
20. apology + es

practice 2

Combine each word with the suffix indicated. Write each new word, following the rules for adding suffixes to words with final *y*. There are eight exceptions in this list.

1. opportunity + es
2. Toby + 's
3. cry + ed
4. mystery + ous
5. satisfy + ed
6. family + 's
7. lazy + est
8. anybody + 's
9. employ + er
10. country + es
11. justify + ed
12. Murphy + s
13. gloomy + ly
14. qualify + ed

15. qualify + ing
16. necessary + ly
17. satisfy + ing
18. Kennedy + s
19. accuracy + es
20. day + ly
21. lonely + ness
22. shy + ness
23. apply + ance
24. lady + es
25. lady + 's
26. weary + ly
27. comply + ance
28. busy + ly
29. identify + es
30. journey + ed

← application →

Proofread the following paragraphs carefully. Then rewrite the paragraphs, correcting any spelling mistakes you have found. There are seven misspellings of words with suffixes.

> The koala, which looks like a toy teddy bear, was first identifyed by explorers as a bear. However, it is a marsupial, that is, a mammal with a pouch for carrying its young.
>
> Koala babys are not developed enough at birth to live outside the mother's pouch, often staying inside the pouch for about six months.
>
> The koala lives in the tops of a few varietys of eucalyptus trees, easyly satisfying its hunger with young eucalyptus leaves and buds. The koala relys solely on this limited diet and seldom drinks water. It is a sluggish animal, sleeping lazyly during the day and moving in an unhurryed way from one eucalyptus tree to another.

53 Suffixes After Final *c*

Words like *picnic* and *shellac* have a final letter *c* that represents a /k/ sound. When certain suffixes are added to words like these, you need to make a special spelling change to remind the reader that the *c* is still pronounced /k/, not /s/.

The letter *c* usually represents an /s/ sound when it comes before the letters *i* and *e*, as in *exer**ci**se* and ***ce**llar.* Because this is true, when you add a suffix beginning with *i* or *e* to words like *picnic* and *shellac,* you must also add the letter *k* after the *c* to indicate the /k/ pronunciation.

picnic /k/ + ing = picnicking /k/, not *picnicing /s/
shellac /k/ + ed = shellacked /k/, not *shellaced /s/

Of course, when the letter *c* represents the /s/ sound before a suffix beginning with *i* or *e,* you don't need to add a *k.*

critic /k/ + ize = criticize /s/
fanatic /k/ + ism = fanaticism /s/

Also, if a letter other than *i* or *e* begins the suffix, no spelling change is needed to indicate that the letter *c* represents the /k/ sound.

historic /k/ + al = historical /k/
economic /k/ + s = economics /k/

practice 1

Combine each word with the suffix indicated, saying each new word aloud. Then write each new word, adding a *k* if it is needed.

1. mimic + ing
2. domestic + ate
3. critic + al
4. frolic + ed
5. characteristic + s
6. basic + ally
7. panic + ing
8. electric + al
9. critic + ism
10. alphabetic + al
11. traffic + ing
12. analytic + ally
13. panic + ed
14. frolic + some
15. symbolic + ally
16. traffic + ed
17. politic + al
18. shellac + ing
19. politic + ize
20. picnic + ed
21. frolic + ing
22. symmetric + al
23. specific + ally
24. mimic + ed

application

Rewrite each sentence, replacing the word or words in heavy type with the word in the list that has a similar

*The asterisk (*) indicates incorrect spelling.

meaning. Use a dictionary if you need help with any of the meanings.

basically	panicked
criticism	picnicking
fanaticism	political
frolicked	shellacked
mimicking	trafficked

1. Everyone in our family enjoys **an outing with eating in the open** during the hot summer months.
2. Ms Sapolski encouraged us to use constructive **evaluation and analysis** when we discussed our reports.
3. After the table had been sanded, it was **coated with shellac** to protect the surface.
4. Mr. LeFevre was pleased at our **precisely imitating** his pronunciation in French class.
5. Heavily **traveled** roads must be cleared of parked cars during rush hours.
6. At recess the children **played and ran happily** in the fallen leaves in the school yard.
7. Although the pilot had to make an emergency landing, no one **became suddenly and overpoweringly frightened.**
8. The student council agreed that Martha's plan for raising money was **fundamentally** sensible.
9. The mayor told us she had always been interested in anything **relating to the conduct of government.**
10. Many who heard the speech were disturbed by the speaker's **behavior of intense, uncritical enthusiasm.**

54 Noun Plurals

Spelling the plural form of nouns is usually easy and causes little trouble.

You can write the plural form of most nouns by simply adding *-s*.

island**s**	muscle**s**	idea**s**
the Tanaka**s**	the Smith**s**	the Wolinski**s**

For nouns that end in *s, sh, ch, x,* and *z,* you write the plural form by adding *–es.*

campus**es**	brush**es**	hoax**es**
cross**es**	search**es**	buzz**es**
the Jones**es**	the Schwartz**es**	the Andros**es**

Some noun plurals can be troublesome, however. The groups of nouns that follow have special rules for forming plurals, depending on how the nouns themselves are spelled.

Final *y* For nouns ending in *y* following a vowel, you just add *–s.*

survey + **s** = survey**s** holiday + **s** = holiday**s**

For nouns ending in *y* following a consonant, you change the *y* to *i* and then add the plural *–es.*

liberty + **es** = libert**ies** diary + **es** = diar**ies**

Final *o* For nouns ending in *o* following a vowel, you just add *–s.*

ratio**s** rodeo**s** studio**s** patio**s** radio**s**

For most nouns ending in *o* following a consonant, you also add *–s.* Notice that these nouns include many musical terms.

Filipino**s**	taco**s**	piano**s**	soprano**s**
Eskimo**s**	alto**s**	cello**s**	concerto**s**
poncho**s**	silo**s**	memo**s**	pimento**s**

But for a few nouns ending in *o* following a consonant, you add *–es.*

lingo**es**	tomato**es**	potato**es**
echo**es**	hero**es**	veto**es**

Finally, you may add either *–s* or *–es* to spell the plurals of a few nouns like *tornado, banjo, cargo, zero,* and *mosquito.*

If you want to be sure of the plural spelling for nouns ending in *o* following a consonant, use a dictionary.

Final *f* and *fe* For some nouns ending in *f* and *fe,* you just add *–s.*

belief**s**	gulf**s**	safe**s**	chief**s**
proof**s**	roof**s**	muff**s**	giraffe**s**

But for a few nouns that have been in the language a long time, you change *f* or *fe* to *v* before adding *–es*.

self + es = sel**ves**	life + es = li**ves**
half + es = hal**ves**	wife + es = wi**ves**
wolf + es = wol**ves**	knife + es = kni**ves**
loaf + es = loa**ves**	shelf + es = shel**ves**
leaf + es = lea**ves**	elf + es = el**ves**
calf + es = cal**ves**	thief + es = thie**ves**

Irregular change A few nouns that have been in the language for a long time have special plural forms. To spell these irregular noun plurals, you must make an internal change in the word or add an ending other than *–s* or *–es*.

woman	—	wom**e**n	mouse	—	mi**c**e
man	—	m**e**n	goose	—	g**ee**se
tooth	—	t**ee**th	ox	—	ox**en**
foot	—	f**ee**t	child	—	child**ren**

No change Some nouns have the same form for singular and plural. Note, for example, these names of animals, fish, and national groups:

deer	Chinese	Swedish
sheep	Sioux	Japanese
moose	Swiss	Spanish
swine	Polish	English
trout	Portuguese	French

practice 1

Write the plural of each of the following words, adding *–s* or *–es*.

1. coach
2. Musso
3. garage
4. trouble
5. dress
6. magazine
7. tax
8. judge
9. opinion
10. bonus
11. topaz
12. batch
13. Finch
14. specimen
15. sneeze
16. box
17. toothbrush
18. choice
19. nickel
20. crash

practice 2

Write the plural of each of the following words, making any changes necessary to form the plural correctly.

1. life
2. cello
3. journey
4. radio
5. belief
6. veto
7. pulley
8. hero
9. grocery
10. tooth
11. taco
12. Chinese
13. emergency
14. shelf
15. allergy
16. woman
17. Filipino
18. activity
19. chimney
20. potato
21. self
22. tendency
23. studio
24. industry
25. deer
26. piano
27. chief
28. ability
29. mouse
30. leaf

application

Each of the following sentences has two singular words in parentheses. After the number of each sentence, write the plural form of the two words.

1. The (Jones) are growing their own (tomato) this summer.
2. (Knife) and (poncho) are essential equipment for serious backpackers.
3. On Wednesday night, the (man) of the square-dance group are preparing dinner for the (woman).
4. Dave announced, "My favorite kinds of fruit are (peach) and (strawberry)."
5. Elsa enjoys (magazine) with science-fiction (story).
6. Growing (cactus) is just one of Jamie's (hobby).
7. Special memorial services will be observed in the (church) and (synagogue) next week.
8. Environmentalists insist that (wolf) as well as (sheep) are important in preserving the balance of nature.
9. Last week the union leaders made several (speech) about current (wage).
10. The (Cowichan Indian) are sending a representative to Victoria to discuss educational (policy).

55 Words with Prefixes

When you add a prefix to a word to form a new word, you usually make no change in the spelling of the prefix or in the spelling of the word.

anti + freeze = antifreeze
un + aware = unaware
dis + regard = disregard
pre + pay = prepay

Sometimes, the last letter of a prefix and the first letter of the word it joins are the same—for example, the letter *n* in *un + necessary*. Notice that the double letters remain in the new word, ***unn**ecessary*.

inter + racial = inte**rr**acial
trans + ship = tran**ss**hip
sub + basement = su**bb**asement
mis + spell = mi**ss**pell

Double letters may also result when you use the prefixes *im–, il–, ir–,* and *in–,* all with the meaning "not."

im + moral = i**mm**oral
ir + regular = i**rr**egular
il + logical = i**ll**ogical
in + numerable = i**nn**umerable

Sometimes the double letter is a vowel—for example, the letter *e* in *re + elect*. In such a case, you may find the new word spelled *reelect,* without change. Or, you may find it spelled *re-elect,* with a hyphen to separate the two vowel letters. Because there may be alternative spellings for such words, you should check your dictionary and use the spelling recommended by your teacher.

Occasionally, you will need to use a hyphen after a prefix to distinguish a word you want to use from another word with an identical spelling but with an entirely different meaning.

coop (a small enclosure)
co-op (a cooperative)
recollect (to remember)
re-collect (to collect again)

You should always use a hyphen to join a prefix to a proper noun or a proper adjective.

mid-Atlantic pre-Islamic inter-American

Finally, use a hyphen to join the prefix *self–* to a word.

self-disciplined self-esteem self-supporting

practice 1

Combine each prefix with the word indicated. Write each new word, making any changes necessary to spell it correctly.

1. dis + appear
2. hyper + active
3. un + natural
4. in + capable
5. post + Renaissance
6. sub + zero
7. inter + related
8. in + efficient
9. non + violent
10. ir + responsible
11. in + justice
12. self + conscious
13. anti + smog
14. re + commend
15. in + accurate
16. un + usual
17. self + sufficient
18. im + balance
19. re + cover (to cover again)
20. in + dependent

← application →

Rewrite each sentence, replacing the word or words in heavy type with the word in the list that has a similar meaning. Use a dictionary if you need help with any of the meanings.

antiwar	immature	resigned
tricentennial	irrelevant	re-signed
disagree	misstated	self-respect
illegal	pre-Confederation	semiannually
illiterate	prepackaged	semicircle

1. No one who is **not able to read or write** will be considered for the job.
2. Paul asked, "Is it **unlawful** to change lanes without signaling?"
3. One should always have **a proper respect for oneself as a human being.**
4. Because commuting to and from work took three hours a day, Jody finally **quit.**
5. The speaker's presentation of the lawyer's view was **not correctly expressed.**

6. The Hudson's Bay Company celebrated its **three-hundredth anniversary** in 1970.
7. Since witnesses were not present the first time, the will had to be **signed again.**
8. Many consumers object to buying vegetables that are **packaged before being offered for sale.**
9. Mark suggested that club elections be held **every six months, or twice a year.**
10. The judge ruled that the lawyer's interruption was **not relating to the matter at hand.**
11. Claire and Andy remain good friends, although they often **differ in opinion** about politics.
12. As the music began, the folk dancers joined hands and formed a **half circle.**
13. The guide didn't know in what period the jewelry had been made, but she thought it was **before the time of Confederation.**
14. "In foreign policy," said the candidate, "I am firmly **against armed conflict.**"
15. Plants and shrubs that are **not fully grown or developed** must be protected against the winter cold.

56 Words That Sound Alike

Some words sound alike but have different spellings and different meanings, like *brake* and *break.* Since the sound of such words gives no clue to their spelling, choosing the correct spelling can sometimes be a problem. It helps if you can remember something that links the spelling of a word to its meaning—for example:

> *Stationery* is for writing letters; *stationary* means "standing still."

Here are several of these troublesome word groups. Try to connect spellings with meanings so that the words won't be troublesome any longer.

Their, There, They're

Their means "belonging to them."

> Anne and Jeff left **their** bikes in the garage.

There may mean "at that place." Or, it may function as a sentence opener when the real subject follows the verb.

The baking dishes are over **there** by the cabinet.

There was a telephone call for you last night.

They're is the contraction of *they are.*

The band members are ready; **they're** waiting for you.

Your, You're

Your means "belonging to you."

I hope **your** sore throat is better soon.

You're is the contraction of *you are.*

I hope **you're** feeling better today.

Its, It's

Its means "belonging to it."

Indonesia gained **its** independence in 1949.

It's is a contraction of *it is* or *it has.*

It's good to see you again. [*it is*]

It's been a long time since we've met. [*it has*]

Whose, Who's

Whose means "belonging to whom."

Whose pen is this?

Who's is a contraction of *who is* or *who has.*

Who's there? [*who is*]

Who's been using my transistor radio? [*who has*]

Compliment, Complement

Compliment means "an expression of esteem or admiration."

The speaker paid a **compliment** to her host.

Complement means "something that completes."

In team teaching, the work of one teacher **complements** the work of the other.

All ready, Already

All ready is a phrase meaning "completely ready."

Susan was **all ready** by lunchtime.

Already is a single word meaning "before a given time."

They were **already** finished by lunchtime.

All together, Altogether

All together, a phrase, means "everyone together."

The guests are **all together** in the main dining room.

Altogether, a single word, means "in all" or "wholly."

Altogether, there were twelve men and nine women.

Understanding him is an **altogether** different problem.

Principal, Principle

Principal can be a noun meaning "a head person" or an adjective meaning "most important."

The **principal** of our school plays the drums. [noun]

Our **principal** reason was to see you. [adjective]

Principle means "a rule or code of conduct." Think of the letters *le,* which end both *princip**le*** and *ru**le.***

The **principle** underlying this activity is not clear.

Passed, Past

Passed is the past tense form of *pass.* The *–ed* ending signals that it is a verb form.

Joe **passed** his chemistry exam with a high *B.* [verb]

Past can function as a noun, a preposition, or an adjective, depending on its use in a specific sentence.

Grandmother likes to tell us about the **past.** [noun]

We drove **past** the new town hall today. [preposition]

Sam has studied guitar for the **past** month. [adjective]

practice 1

For each sentence, decide which spelling in parentheses represents the word that correctly fits the blank. After the number of the sentence, write the word.

1. ___ is probably a perfectly natural reason for such strange behavior. (There, Their, They're)
2. Dad just ___ the turnoff by mistake. (passed, past)
3. "___ a pleasure to see you," said Ms. Wyatt. (Its, It's)
4. I think Elva has ___ eaten dinner. (all ready, already)
5. Jean's sense of humor ___ her serious attitude toward studying. (compliments, complements)

6. ___, about thirty people couldn't get tickets. (All together, Altogether)
7. For the ___ two semesters, Marilyn has been taking woodworking as an elective. (passed, past)
8. The committee submitted ___ report today. (its, it's)
9. I hope ___ bus won't be late. (their, there, they're)
10. I hope ___ on time today. (their, there, they're)
11. Irving was pleased to receive a ___ from his teacher after his report. (compliment, complement)
12. Go ___ the station and turn left. (past, passed)
13. Do you know ___ coming this evening? (whose, who's)
14. If ___ going to the library, would you take my books back? (your, you're)
15. Both candidates were known as people of high ideals and firm ___. (principals, principles)

← application →

Write nineteen sentences of your own, using one of the following words or phrases in each sentence.

1. its
2. it's
3. whose
4. who's
5. your
6. you're
7. their
8. there
9. they're
10. past
11. passed
12. already
13. all ready
14. principal
15. principle
16. compliment
17. complement
18. all together
19. altogether

57 Easily Confused Spellings

Some words, like *quiet* and *quite,* sound almost alike or look almost alike. However, their spellings and their meanings are different. If you are not careful, you may use the wrong spelling.

Here are several word pairs containing words whose spellings are often confused. Try to connect the spelling of each word with the meaning and pronunciation you intend.

Accept, Except

Accept, a verb, means "to receive."

Rahim will **accept** the award for the team.

Except, a preposition, means "excluding," "leaving out."

All the books **except** those in boxes were damaged.

Thorough, Through

Thorough has two syllables. It is a modifier meaning "complete," "careful," "detailed."

Dr. Fong recommended that she have a **thorough** physical examination.

Through has one syllable and rhymes with *threw*. It may indicate passage from one end of something to the other, or it may mean "finished."

The nail went completely **through** the board.

We are finally **through** with that committee work.

Proceed, Precede

Proceed means "to go forward" or "to continue."

Proceed with your work; don't let me bother you.

Precede means "to go in front of" or "to come before."

The luncheon will **precede** the presentation of awards.

Formerly, Formally

Formerly means "previously" or "in earlier times."

Formerly, only seniors could take consumer education.

Formally means "in a formal, dignified manner."

Ms Kolinek dressed **formally** for the dinner party.

Advice, Advise

Advice is a noun. It rhymes with the noun *rice.*

I wish I had some good **advice** to offer you.

Advise is a verb. It rhymes with the verb *rise.*

Carl didn't **advise** me about which courses to take.

Personal, Personnel

Personal means "relating or belonging to a person."

These are my **personal** papers; don't disturb them.

Personnel means "a group of people, often employees."

All interviews are held at the **personnel** office.

Allusion, Illusion

Allusion means "an indirect reference."

Did anyone understand the **allusion** to *Catch 22?*

Illusion means "something that misleads or deceives."

The fog gave us the **illusion** that the road led right into the river.

Conscience, Conscious

Conscience, a noun, means "a sense of right and wrong."

Dave's **conscience** made him stay to help the others.

Conscious, an adjective, means "aware" or "awake."

Sue was **conscious** of the change in Janie's attitude.

Alan was stunned by the blow but still **conscious**

Later, Latter

Later means "after the usual time."

Should I take the earlier train or the **later** one?

Latter means "the second of two things mentioned."

Glue or tape will do; the **latter** is easier to use.

Moral, Morale

Moral means "relating to ideas of right or wrong" or "conforming to a standard of right behavior."

Steve felt a **moral** obligation to help, although he had no legal responsibility toward them.

Morale means "the sense of well-being of a person or a group."

The team's **morale** improved after the coach's talk.

practice 1

For each of the following sentences, decide which spelling in parentheses represents the word that correctly fits the blank. After the number of each sentence, write the word.

1. With only one lane open, traffic ___ slowly across the bridge under repair. (preceded, proceeded)
2. Matt ate everything ___ the pie. (except, accept)
3. Our physical education program was ___ much smaller than it is now. (formerly, formally)

4. This essay is especially difficult because of the number of ___ to literary figures. (illusions, allusions)
5. The driver of the pickup was stopped for going ___ a red light. (thorough, through)
6. Dee's ___ was to review carefully. (advise, advice)
7. The graduates will ___ the other classes marching into the assembly hall. (proceed, precede)
8. ___ , all the tickets were sold. (Latter, Later)
9. ___ is low since Mr. Morelli resigned. (Moral, Morale)
10. The new ___ policy regarding overtime came as a surprise to many employees. (personnel, personal)
11. The detective made a ___ search of the room for fingerprints. (through, thorough)
12. By having a local anesthetic, Rita remained ___ while her tooth was pulled. (conscience, conscious)
13. The heated air over the hot pavement created an ___ of a pool of water. (allusion, illusion)
14. You can take the old highway or the bypass; the ___ is fifteen minutes faster. (later, latter)
15. The sign over the coatrack read Not Responsible for ___ Property. (Personnel, Personal)

application

Write twenty sentences of your own, using one of the following words in each sentence.

1. accept
2. except
3. proceed
4. precede
5. moral
6. morale
7. advice
8. advise
9. later
10. latter
11. personal
12. personnel
13. allusion
14. illusion
15. conscious
16. conscience
17. through
18. thorough
19. formerly
20. formally

Building Vocabulary〉

58 Context Clues to Word Meanings

When you come across an unfamiliar word in your reading, don't just skip over it. Think about the context in which the unfamiliar word appears—that is, the other words that surround it. See what clues the meaning of the other words in the sentence may provide to the meaning of the unfamiliar word.

First, consider the general subject matter of the entire sentence. Is it sports, food, politics, music? Knowing even this much helps you narrow the range of meaning of the unfamiliar word. For example, what is the general subject matter in the example sentence that follows? And what does *pseudonym* mean?

> Some women writers in the 1800s used a man's name as a **pseudonym** so that readers would not know that the author was a woman.

You might decide that the general subject is writing, or writers, or perhaps women writers. You still don't know what *pseudonym* means, but it probably has something to do with writing.

Next, look at other words in the sentence that may explain the meaning of the unfamiliar word even further. In the example sentence, the words *women writers . . . used a man's name* give you additional clues that *pseudonym* must have something to do with using a false name when writing.

So, you might decide that *pseudonym* means "a false name" or "a pen name." Of course, you can't be positive. But you can make an educated guess that should help you read through to the end of the paragraph.

Sometimes, a sentence with an unfamiliar word doesn't provide helpful clues to the meaning of the word. In such cases, other sentences in the paragraph may help. Here is an example in which the meaning of an unfamiliar word is cleared up in a following sentence:

> My first attempt at skiing was a **fiasco** Not only did I fall, but I couldn't even get back up on my skis without help.

The first sentence tells you that attempting to ski is the general subject matter. But what, in this context, is a *fiasco?* The second sentence gives you clues with words like *fall, couldn't get back up . . . without help.* Clearly, the first attempt at skiing was not even partly successful. So, you could reasonably decide that *fiasco* means "not successful" or "a complete failure."

You can see that you may be able to figure out the meaning of an unfamiliar word from its context—the sentence in which the word appears or other sentences in the paragraph. Remember to look for two things—the general subject matter and the meaning of surrounding words. But if you are still in doubt about the meaning of a word when you finish reading the paragraph, reach for your dictionary.

practice 1

In each of the following sentences, use context clues to choose the meaning that is closest to the meaning of the word in heavy type. After the number of the sentence, write the word in heavy type and the meaning you have chosen.

1. After the strong winds of the week before, everyone welcomed the softly blowing **zephyr.**
   a. a heavy rain
   b. a gentle breeze
   c. a hot spell
   d. a hurricane
2. Connecting the two land masses is a narrow **isthmus** that separates the Atlantic and Pacific oceans.
   a. a strip of land with water on two sides
   b. a cliff
   c. an island with water on all sides
   d. a valley
3. Cattle, deer, and many bird are **herbivorous.** They cover large areas in search of grasses and seeds.
   a. wild
   b. meat eating
   c. plant eating
   d. tame
4. Pablo Casals played the cello so expertly that he was widely recognized as a **virtuoso.**
   a. a spectator
   b. an amateur musician
   c. a singer
   d. a skilled musician
5. It was hard to ignore the **strident** voices of the protesters as they shouted their demands.
   a. harsh, insistent
   b. pleasant
   c. soft, gentle
   d. melodic

6. There was a **hiatus** in the hostilities. Both sides had agreed to a temporary cease-fire.
   a. a gap or break
   b. a continuation
   c. a beginning
   d. a final ending
7. The Jacksons combined colonial, victorian, and modern furniture, achieving a pleasing **eclectic** style.
   a. unattractive
   b. composed of the best from one source
   c. disagreeable
   d. composed of the best from various sources
8. Although neither one said anything, agreement was **implicit** in the smiles and handshakes of the two men.
   a. stated, plainly expressed
   b. doubtful
   c. implied, not directly expressed
   d. broken
9. Mrs. Jensen never used more words than she needed. In fact, she was **laconic** even with her best friends.
   a. sarcastic
   b. using many words
   c. untruthful
   d. using few words
10. My father, who grew up in a large, poor family, learned as a boy how to save and to be **frugal.**
   a. reckless
   b. thrifty
   c. wasteful
   d. indifferent

practice 2

Use context clues in each of the following sentences to pick the word in heavy type that best completes each sentence. Write the complete sentence.

scrutinize: to examine closely
voracious: greedy, having a huge appetite
affable: friendly, gracious, at ease
litigation: a lawsuit
upbraid: to scold angrily
regale: to entertain, to give pleasure or amusement
anthology: a collection of literary pieces
lethargic: lacking in alertness or activity; sluggish

1. The lawyers announced that the disagreement would be settled out of court, without ___ .
2. Tim's ___ behavior made his parents suspect that he was not getting enough sleep.

3. The committee members promised to ___ the school expansion plans carefully.
4. Ms. Larson was ___ and courteous as she greeted the newcomers.
5. Mr. O'Brien liked to ___ his friends with stories about his years on the stage.

← application →

Pick ten of the words in heavy type in this lesson. Then write ten sentences of your own, using one of the words in each sentence.

59 Similar Words as Meaning Clues

One way to arrive at the meaning of an unfamiliar word in a sentence is to look for other words whose meaning you already know. See if the familiar words can be linked to the unfamiliar word in such a way as to provide clues to its meaning.

For example, the unfamiliar word may be one of several words in a series of synonyms—words with similar meanings. If you know the meaning of at least one of the other words in the series, you should be able to guess at the meaning of the unfamiliar word. What words give you clues to the meaning of *staid* in this sentence?

> The **staid**, serious, and self-restrained manner of the judge always had a steadying effect on people.

Because the unfamiliar word *staid* is linked with *serious* and *self-restrained* in a series, you could reasonably assume that it is close to these two words in meaning. So, you might decide that *staid* must mean "serious, composed, steady."

Sometimes an unfamiliar word is linked to a familiar word by *and.* Suppose you don't know the meaning of the word *defraud.* What clue to its meaning can you find in this sentence?

> The store owner's attempt to deceive and **defraud** his customers put him in serious trouble.

Because *defraud* and *deceive* are joined by *and,* you can assume that the two words are probably close in meaning. So, you might decide that *defraud* must mean something like "to deprive someone of something by deceit; to cheat."

Sometimes a word group, rather than a single word, has the same meaning as an unfamiliar word in the same sentence. What clue do you find to the meaning of *officiously* in the word group that follows it?

> Diane thought her neighbor acted **officiously**, giving advice that was not asked for and not needed.

The word group *giving advice that was not asked for and not needed* describes how the neighbor acted. Since *officiously* also describes how the neighbor acted, you can assume that it means "in an interfering, meddlesome manner."

Remember, to make an educated guess about the meaning of an unfamiliar word in a sentence, look for other words you know that probably have a similar meaning. To make certain your guess is correct, check the word in your dictionary.

practice 1

In each sentence look for the word or words that probably have a similar meaning to the word in heavy type. Then pick from the four choices the word or words closest in meaning to the word in heavy type. After the number of each sentence, write the word in heavy type and the meaning you have chosen.

1. The children, **restive** and uneasy, were still trying to challenge the new bedtime rules.
   - a. happy, unconcerned
   - b. quietly accepting, easy to manage
   - c. calm, relaxed
   - d. stubbornly resisting, hard to manage
2. Unfortunately, the meal was dull and **insipid,** and the conversation was the same.
   - a. exciting, lively
   - b. flat, flavorless
   - c. spicy, hot
   - d. interesting, flavorful

3. The treasurer's report was short, simple, and **succinct.**
   a. brief and clear
   b. vague and rambling
   c. complicated and wordy
   d. lengthy and confusing
4. The coach **reproached** the team, expressing her displeasure at the way they played.
   a. praised
   b. criticized
   c. congratulated
   d. joked with
5. Lew was **adamant,** firm, and completely unyielding in opposing our suggestions.
   a. willing to change
   b. relaxed
   c. unwilling to change
   d. undecided
6. The wheat crop was **copious,** yielding an abundance of low-priced wheat products.
   a. insufficient
   b. expensive
   c. plentiful
   d. delicious
7. The conservation group will remain **vigilant,** checking carefully for any signs of water pollution.
   a. indifferent
   b. unconcerned
   c. contented
   d. watchful
8. The committee promised to examine the situation carefully, to **probe** any suggestion of misconduct.
   a. ignore completely
   b. investigate thoroughly
   c. encourage fully
   d. enjoy briefly
9. The watchman was considered **remiss** and careless for sleeping while at work.
   a. neglectful of duty
   b. thoughtful
   c. concerned
   d. attentive to duty
10. The clown's **ludicrous,** comical behavior brought waves of laughter from the crowd.
    a. sad
    b. angry
    c. ridiculous
    d. serious

practice 2

Answer each of the following questions according to your understanding of the word in heavy type. After the number of each question, write *yes* or *no.* Use a dictionary if you need help with any of the meanings.

1. Would you want to go to a movie that had an **insipid** plot?
2. Should a crossing guard at an elementary school be **vigilant?**
3. Does a person who speaks **succinctly** take a lot of time and use many words to say something?
4. Would you expect a doctor to **probe** for symptoms of illness in a patient?
5. If you made a **ludicrous** statement, would you expect your friends to laugh?
6. If a student takes **copious** notes, would you expect her to use a lot of paper?
7. Would a person who was **remiss** in his duties make a good baby-sitter?
8. Would you expect students who were **restive** about a regulation to want to speak to the principal about it?
9. Is it easy to talk with people who are always **adamant** in their opinions?
10. If your brother borrowed your bike and left it out in the rain, would you **reproach** him?

application

Pick ten of the words in heavy type in this lesson. Then write ten sentences of your own, using one of the words in each sentence.

60 Contrasting Words as Meaning Clues

Finding clues to what an unfamiliar word does *not* mean can sometimes be the best way to determine what it *does* mean. If you come across an unfamiliar word in a sentence, see if there is a familiar word nearby that contrasts with it, a word with an opposite meaning. Since you know what the familiar word means, the contrast will give you a clue to the meaning of the unfamiliar word.

In the following sentence, what clues are there to the meaning of *ornate?*

> The restaurant was **ornate**, not plain and simple as we had expected.

The word *ornate* is contrasted with the words *plain* and *simple*. The word that signals the contrast is *not*. If *plain* and *simple* both mean "without decoration," then *ornate* probably means the opposite. You could assume that *ornate* must mean "heavily decorated," "elaborate."

Besides *not*, other words that may signal contrast are *although, however, but,* and *nevertheless*. What word signals contrast in the example sentence that follows? How can you figure out what the word *dearth* means?

> Although the hikers started out in a happy and talkative mood, there was a **dearth** of laughter and conversation when they returned twelve hours later.

In this sentence, the word *although* signals a contrast between the two moods of the hikers. They started out happy and talkative and returned in the opposite mood, with a *dearth* of laughter and conversation. So, you could decide that *dearth* must mean "an absence or scarcity."

Sometimes when there is a contrast in a sentence, the contrasting words may not be exactly opposite in meaning. For instance, in the following sentence, does *tepid* mean "cold," the exact opposite of *hot?*

> At lunchtime, our soup was **tepid**, not hot.

The word *tepid* is not the exact opposite of *hot*. It does not mean "cold," but "moderately warm," "lukewarm." Unfortunately, you can't tell from the example sentence what the exact meaning of *tepid* is. The most you can tell is that something *tepid* is "not hot." You need your dictionary to find out the exact degree of heat *tepid* represents.

Finding words with contrasting meaning in a sentence can help you make an educated guess about the meaning of an unfamiliar word. You should remember, however, that contrast clues may not tell you exactly what a word means. If you are still in doubt about an unfamiliar word in a sentence after considering the words with contrasting meaning, use your dictionary to pin down the exact meaning.

practice 1

Use the contrast clues in each sentence to determine the meaning of the word in heavy type. Then pick from the four choices the word or words closest in meaning to the word in heavy type. Write the word in heavy type and the meaning you have chosen.

1. Although campers are free to choose most activities, the swimming classes are **compulsory.**

   a. not required — c. easy
   b. required — d. tiring

2. The foreman said he was **flexible;** nevertheless, he refused to consider a new work schedule.

   a. very unhappy — c. always impatient
   b. rigidly firm — d. capable of change

3. The young lawyer began by **advocating** passage of the new bill, but she ended by arguing against it.

   a. pleading for — c. not supporting
   b. disapproving of — d. resisting

4. Ben was **abstemious** at dinner last night; however, today he returned to his habit of eating everything in sight.

   a. hungry — c. restrained
   b. greedy — d. sloppy

5. Although Joe lacked formal training, his experience made him **competent** to handle the job.

   a. qualified — c. unwilling
   b. eager — d. unfit

6. Ms. Gibson liked to **improvise,** not to make precise arrangements or to prepare her remarks in advance.

   a. plan carefully — c. do without planning
   b. joke — d. sing

7. A **lucrative** summer job is best; nevertheless, even work with little or no pay is useful experience.

   a. easy — c. enjoyable
   b. interesting — d. profitable

8. Although Marie worked hard to give an excellent report, her lack of information made it only **mediocre.**

   a. of high quality — c. amusing in tone
   b. of ordinary quality — d. angry in tone

9. Art is **pessimistic** about tomorrow's game, but the other team members remain hopeful.

   a. expecting a bad result — c. in a happy mood
   b. expecting a good result — d. in a calm mood

10. Runners with **stamina** finished easily; however, others tired quickly and fell out of the race.
    a. loss of appetite
    b. poor training
    c. no money
    d. staying power

practice 2

In each of the following sentences, find the words that contrast in meaning with one of the words in heavy type that follow. Then write the sentence, adding the word in heavy type that best completes it.

1. Joanie was trying to __ the preparations for the party; nevertheless, her trying to do everything herself only served to make things harder.
2. The biologist warned, "Although the present generation is safe, we must act now to protect __."
3. At first Mark felt __ about missing the game, but later the coach's visit cheered him up.
4. Rita wants __ proof of the existence of flying saucers; however, she has never touched or even seen one.
5. Mr. Ryan was always __ when he counseled students; he didn't beat around the bush.

evasive: indirect, avoiding a straightforward answer
dejected: in a low mood, depressed
posterity: descendants, future generations
imperceptible: not capable of being seen or touched
elated: in a joyous mood, marked by high spirits
facilitate: to make easier
contemporaries: people living at the same time
complicate: to make complex or difficult
candid: straightforward, honest, frank
tangible: real, capable of being seen or touched

← application →

Pick ten of the words in heavy type in this lesson. Then write ten sentences of your own, using one of the words in each sentence.

61 Explanations as Meaning Clues

Sometimes a writer must use a word that may be unfamiliar to the general reader. In such cases, a careful writer will supply an explanation to make certain the reader understands the word or the special way in which it is being used. It is up to the careful reader to recognize and apply the explanation as a clue to word meaning. For instance, what tells you the meaning of *cinch* in this sentence?

> As we saddled our horses, Mr. Gomez told us to make the **cinch**, or bellyband, as tight as we could.

The explanatory phrase *or bellyband* acts as an instant definition for the word *cinch.* If the word has several meanings, as *cinch* does, such an explanation can also tell you which meaning the writer intends.

There are several ways to include explanations in sentences. Occasionally, a writer will tell you exactly what meaning is intended.

> By ***predator***, I mean any animal that hunts and kills other animals for food.

> The word ***reconcile*** is used here in the sense of "restore to friendship or harmony."

More often, a phrase like *that is, in other words,* or *for example* will precede an explanation.

> Citizens responded to the news with **apathy**—that is, **with complete indifference**

Sometimes the explanation has no introductory words but is simply set off from the unfamiliar word by commas or dashes.

> The guide warned that a **tyro, a beginner in learning,** should not attempt rock climbing alone.

> My interest in education reached its **nadir—the lowest point—** when I had to repeat algebra.

Sometimes the explanation is a word or phrase preceded by the word *or.*

> Mr. Rubin had a tendency to **digress**, or **get off the subject**, whenever he talked about the 1960s.

And sometimes the explanatory word or phrase will come first, and the unfamiliar word will follow.

> The lawyer's manner struck us as being both **secret and sly**—in short, **furtive**

Explanations like these are included by the writer as aids to understanding. Use them as an easy way to add unfamiliar words to your vocabulary.

practice 1

Each of the following sentences contains a word in heavy type and an explanation of that word. Use the explanation to determine the meaning of the word in heavy type. Then choose from the list the definition that is closest in meaning to the word in heavy type. Write the word and the definition you have chosen.

to shift back and forth
requiring care and hard work
to make something greater
the way a person acts
not neat or careful
firm in resisting or enduring
to prove to be right or reasonable
disturbance, confusion
problem requiring a difficult choice
to make public, disclose

1. The puppy was stubborn and persistent—really **tenacious**—in holding the slipper in its jaws.
2. Our club tried to **augment,** or increase, its funds by raising the monthly dues to one dollar.
3. Tom thought his younger brother was untidy, careless, and sloppy—in short, **slovenly.**
4. The candidate tried to **vindicate** her view on foreign aid—that is, to justify it.
5. By ***turmoil,*** I mean commotion, agitation, and disorder.
6. The temperature **fluctuates** widely—for example, it may vary as much as 30 degrees in one day.

7. The reporter refused to **divulge,** or reveal, the source of his information.
8. The problem seemed to present two solutions, each equally unsatisfactory—a true **dilemma.**
9. Callie found backpacking enjoyable but also **arduous**—a difficult and strenuous pastime.
10. Rachel's **demeanor,** or manner, was straightforward and honest.

practice 2

Each of the following words is followed by two words or phrases. One is close in meaning to the word. The other is opposite in meaning. Write the word and the word or phrase that is close to it in meaning.

1. tyro	beginner; expert
2. arduous	easy; difficult
3. tenacious	weak; firm
4. turmoil	confusion; calm
5. nadir	lowest point; highest point
6. apathy	indifference; concern
7. slovenly	neat and careful; untidy and careless
8. divulge	hide; reveal
9. digress	remain on subject; stray from subject
10. augment	decrease; increase
11. vindicate	defend; blame
12. furtive	sincere; sly
13. dilemma	difficult problem; simple problem
14. reconcile	restore to harmony; destroy harmony
15. fluctuate	remain stable; change

application

Pick ten of the words listed in practice 2. Then write ten sentences of your own, using one of the words in each sentence.

62 The Dictionary and Word Meanings

You can sometimes figure out the general meaning of an unfamiliar word by paying close attention to the other words in the sentence—that is, by using clues in the word's context. But context clues will only help you to make an educated guess.

To get the exact definition of a word, you must look it up in the dictionary. Even then, you'll still need to use context clues. Remember, a word may have several meanings, and the dictionary will list all of them. So, don't stop after reading the first definition. You must find the one meaning that best fits the context in which the word appears.

For example, suppose you read a sentence that contains the word *facade,* and you decide to look the word up in the dictionary. First, think about its context—that is, the meanings of the other words in the sentence.

> Judy tried to hide her disappointment behind a **facade** of carefree indifference.

You'll probably decide that in this context *facade* has something to do with appearance and behavior—how Judy looked and acted. Now, when you check *facade* in the dictionary, you'll be able to decide which of the two definitions you find there best fits the context.

1. the front of a building
2. a false, superficial, or artificial appearance

Clearly, the second meaning of *facade* is the one you want.

Once you have found the meaning that fits a word in a particular context, be sure to read the other meanings, too. Then you'll understand the word when you find it used in other contexts. For example, after checking both meanings of *facade,* suppose you later find this sentence in a description of an old town hall:

> The **facade** was partially replaced and remodeled after a fire in 1937.

Remembering that one meaning of *facade* is "the front of a building," you'll know right away what the writer meant.

Often the context of a word includes not only the surrounding words but also the time during which the word was used or the place it was used. Even a familiar word may have a special meaning that was used during a particular historical period or in a particular country.

The dictionary has special labels that identify special definitions influenced by time or place. It may use the label *obsolete* for a meaning no longer used and *archaic* for a meaning rarely used today. If a word meaning is not common to all English-speaking countries, labels like *British* or *Australian* show where a word carries that meaning.

Suppose, for example, that you are reading the play *King Lear* and find Shakespeare's reference to ". . . mice, and rats, and such small deer, . . ." You might be confused seeing mice and rats referred to as deer. But remember that Shakespeare wrote during the seventeenth century. If you check the dictionary, you'll find that one meaning of *deer* has a special label.

> **deer** 1. *archaic:* animal, especially a small mammal.

So *deer* in this seventeenth-century context simply means any small mammal, not the larger animal with hooves and antlers that we call a deer today.

In using the dictionary to find a specific word meaning, keep in mind the context in which the word appears. And remember that the time or the place in which a word is used may be an important part of its context.

practice 1

Read the two definitions given for each of the following words. Then read the sentences, each of which contains one of the words defined. Decide which meaning is intended in each sentence. After the number of the sentence, write the word and the meaning you have chosen.

dissipate	1. to scatter, to cause to disappear 2. to use up foolishly or heedlessly
nice	1. pleasing, agreeable, kind 2. exact, precise, particular
starve	1. to die of hunger 2. *obsolete:* to die

florid 1. flowery in style, ornate
2. tinged with red, ruddy

nurture 1. to further the development of
2. to nourish, to feed

1. Wild grasses and grains were sown to **nurture** the hungry geese on their journey south.
2. Flood victims may **starve** if food doesn't arrive soon.
3. The brothers **dissipated** the family fortune by gambling.
4. Not everyone enjoys that writer's **florid** prose.
5. Mrs. Nelson likes to point out **nice** distinctions in the meanings of words like *wise* and *knowledgeable.*
6. In Chaucer's *Troilus and Criseyde,* written about 1385, Troilus promises to "live and **starve**" faithful to his love Criseyde.
7. Ann tried to act as if she were having a **nice** time.
8. The strong wind rapidly **dissipated** the clouds of smoke.
9. One purpose of the pilots' survival course was to **nurture** self-reliance and endurance.
10. The farmer had a healthy, **florid** complexion from working outdoors all year round.

practice 2

Read the following sentences, each of which contains a word in heavy type. Look each of these words up in the dictionary, and choose the definition that best fits the word's context. After the number of the sentence, write the word and the meaning you have chosen.

1. Tina's only **fault** is that she doesn't know when to stop joking.
2. The San Andreas **fault** in California caused the great San Francisco earthquake in 1906.
3. Fire inspectors agreed that a highly **volatile** gas had been the cause of the explosion and fire.
4. Debbie's **volatile** disposition made it hard for her to settle down to any project for a long time.
5. Our neighbor, who grew up in England, still refers to the **bonnet** of his car.
6. Sandy found that an old-fashioned **bonnet** gave good protection from the hot summer sun.

7. The defendant was declared **lucid** and able to stand trial.
8. We understood better after Al's clear, **lucid** statement.
9. It's hard to **discriminate** between these shades of blue.
10. Single people felt the tax **discriminated** against them since it favored married people unfairly.

← application →

Write ten sentences of your own, using any of the words in heavy type in this lesson. You may use the same word twice, if you use a different meaning of the word each time.

63 Form Clues to Word Meanings

Often you can guess the meaning of a word by its context—that is, by familiar words that surround it. What clues can you find to the meaning of *tack* in this sentence?

> Saddles, bridles, and other **tack** were kept in a small room behind the stable.

You recognize *saddles* and *bridles* as familiar words for equipment used with horses. You see that the words *and other* connect *tack* to *saddles* and *bridles.* So you decide that the word *tack* must mean horse equipment, too.

But what do you do if an unfamiliar word is standing alone or if its context gives you no help? In these cases, the *form* of the word—how the word is put together—will often give you clues to its meaning.

To find meaning clues in the form a word takes, you must break the word down into its parts. Some words, like *work,* have only a single part. But many other words, like *unworkable,* are combinations of word parts. If you can recognize these word parts, if you know their meanings, then you can make educated guesses about the meanings of unfamiliar words in which these same parts appear.

The Base The part of the word that carries the principal meaning, the core of the word, is called the base. The base of *un**work**able* is *work.* Some bases, like *work,* can stand by themselves as words. Many others, however, must be combined with at least one other word part before they make sense as words. For example, *–ceive* is the base of the word *receive.* Bases like *–ceive* are often difficult to recognize because they are word parts borrowed from Latin and Greek.

The Prefix The part of the word that comes before the base is called the prefix. The prefix in ***un**workable* is *un–*. Because prefixes have meanings of their own, they change the meanings of the bases to which they are attached. Since *un–* means "not," ***un**workable* must mean "not workable." Being able to recognize prefixes can help you figure out the meanings of unfamiliar words.

You probably already know the meanings of many prefixes.

impossible = **not** possible

reheat = heat **again**

antiaircraft = **against** aircraft

predetermined = determined **before**

semicircle = **half** circle

subhuman = **under, below** the human

The Suffix The part of the word that comes after the base is called the suffix. The suffix in *unwork**able*** is *–able.* Suffixes also have meanings of their own, and so they, too, change the meanings of the bases to which they are attached. Since *–able* means "capable of," anything that is *workable* is capable of working or of being worked. Being able to recognize suffixes can help you figure out the meanings of unfamiliar words.

You probably know the meanings of many suffixes already.

connect**or** = **thing that** connects

chem**ist** = **one who** knows chemistry

major**ette** = **little, feminine** drum major

repair**able** = **able** to be repaired

hope**ful** = **full of** hope

beaut**ify** = **to make** beautiful

A good way to guess at the meaning of an unfamiliar word is to see how it is put together. Look at the word carefully. Can it be divided into prefix, base, and suffix? If it can, what clues to the meaning of the whole word do the meanings of its parts give you? You'll still need to use the dictionary sometimes, but not as often if you learn to recognize basic word parts.

practice 1

Rewrite each word. If the word has a prefix, circle the prefix. If you can guess what the prefix means, write your guess after the word. If the word has no prefix, write *no prefix* after the word.

1. regressive
2. semitone
3. ocean
4. indecisive
5. regurgitate
6. incredible
7. entreat
8. dissonant
9. slothful
10. disassemble
11. subservient
12. preeminent
13. antithesis
14. unmitigated
15. financier

practice 2

Rewrite each word. If the word has a suffix, circle the suffix. If you can guess what the suffix means, write your guess after the word. If the word has no suffix, write *no suffix* after the word.

1. arbiter
2. wrathful
3. apologist
4. amplify
5. exclude
6. culpable
7. incongruous
8. defection
9. sensuous
10. vengeful
11. edible
12. epoch
13. supplicate
14. voracious
15. gauntlet

application

Here are lists of prefixes and suffixes:

Prefixes		*Suffixes*	
anti-	against	-et	little
over-	too much	-or	one who
re-	again	-ous	with, having
sub-	under, below	-ize	to make, become
un-	not	-ist	person who
in-	not, in	-ible	able

Using these prefixes or suffixes, write a single word that fits each of the following definitions. As a hint, the correct base is printed in heavy type. Check spellings in the dictionary whenever you are in doubt.

1. one who **prevaricat**es
2. not having **harmony**
3. a little **isl**and
4. against the **establishment**
5. not able to be **flex**ed
6. to make into an **anthology**y
7. the part under the **stratum**
8. a person who knows **physiolog**y
9. not having **gener**osity
10. not able to be **defen**ded
11. to make **vital** again
12. not having **preten**se
13. to make too **sensit**ive
14. to make **legitim**ate
15. person who is against **femin**ism

64 Word Parts as Meaning Clues

It's natural to look at an unfamiliar word in a sentence as if it were an indivisible whole. But this doesn't help much in figuring out what the word means, especially if it's long. Often a long word is a combination of smaller parts, each part contributing something to the meaning of the whole word. Here's a word divided into its parts:

PREFIX → **pre-** BASE → **caution** SUFFIX → **-ary**

When you come across an unfamiliar word, like *precautionary,* try breaking it up into parts you can recognize. If you know *caution* means "carefulness," *pre–* means "before," and *–ary* means "something connected with," you could guess that *precautionary* might mean "something connected with carefulness beforehand." Even though this is not the exact dictionary definition, it should help you read the sentence with better understanding.

Only **precautionary** steps like boiling all drinking water kept the epidemic from spreading.

To get at the meaning of an unfamiliar word, see if it can be divided into parts—prefix, base, and suffix. First locate and try to define its base, the main part of the word. Then consider the meanings of any prefixes that come before the base or suffixes that come after it. If the word still makes no sense, consult a dictionary.

Dividing a word into parts isn't always a foolproof method for defining it. For one thing, words often change in meaning as time goes by. Suppose you know that *–ster* means "one who." Breaking *spin**ster*** into its parts will give you its original meaning, "one who spins." But today *spinster* means "an older, unmarried woman."

Another problem is that the literal meaning of the *parts* of a word may not add up to the meaning of the whole word.

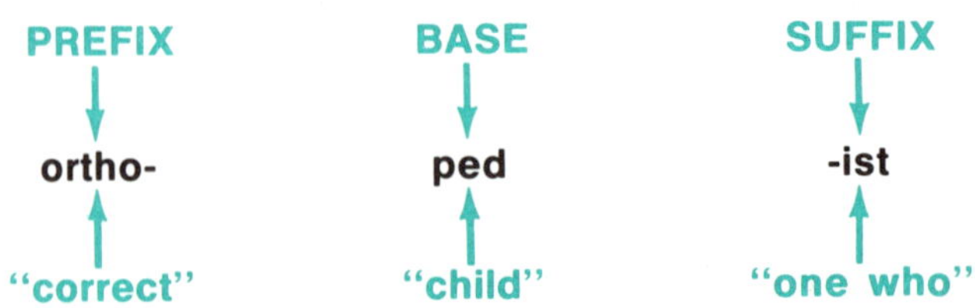

You might assume an orthopedist is a "person who corrects children." But an orthopedist is a doctor whose specialty is correcting bone deformities. If all you had to go on in defining *orthopedist* were the separate meanings of its parts, you wouldn't be able to figure out its real meaning.

A third problem may be recognizing the parts of a word. You'll recognize some bases immediately. In *replay,* for example, you find the familiar word *play.* You probably know that the prefix *re–* means *again.* So you can guess the meaning of *replay* easily. But what about a word like *reduce?* Does the base *–duce* mean anything to you by itself? Probably not, because *–duce* is not a word in English. It's a base borrowed from the Latin word *ducere.* Since *–duce* can't stand alone, it must be bound to another word part, like the prefix *re–.*

Because of problems like these, you can't always count on dividing a word into parts as a method for discovering its meaning. But when the dictionary isn't available, this method will often help you make an educated guess about the meaning of an unfamiliar word.

Lists of some common bases, prefixes, and suffixes follow. Many of them come originally from Latin and Greek words. Most of the bases are like *–duce*. That is, they can't stand alone as English words. Learn the meanings of as many word parts from the following lists as you can.

It is estimated that if you learn just these fourteen groups of bases, you'll be able to recognize over 100,000 words.

Base	*Meaning*	*Example*
cept, ceive, cap, capt, cip, ceit	take, seize	re**ceive** **capt**ure
duc, duct	lead, make	con**duct**
fac, fect, fic, fact	do, make	de**fect** **fact**ory
fer	bear, carry	of**fer**
graph, gram	write	auto**graph**
log, logy	word, study	ana**logy**
mit, miss	send	e**mit**
plic, ply, plex	fold, bend, twist	com**plic**ated
pon, pos	place, put	post**pon**e
scrib, script	write	de**scrib**e
sist, sta	stand, endure	per**sist**
spect, spec, spic	look	con**spic**uous
tain, ten, tin	hold, have	re**tain**
tend, tens, tent	stretch	ex**tens**ion

Here are some more bases that often occur in English:

Base	*Meaning*	*Example*
anthrop	man	**anthrop**ology
arch, archi	rule, govern, first in importance	**arch**bishop mon**arch**
cogn	know	re**cogn**ize
cred	belief, trust	**cred**itable
jud	judge	pre**jud**ice
psych	mind, soul	**psych**ology
soph	wise, wisdom	**soph**isticated
vol	wish	bene**vol**ent

These are some prefixes used often in English words:

Prefix	*Meaning*	*Example*
a–	lacking, without	**a**moral
anti–	against	**anti**war
bi–	two	**bi**centennial
circum–	around	**circum**vent
contra–, contro–	against	**contro**versy
de–	away, from, off, down	**de**cline **de**duce
dis–, dif–	away, off, opposing	**dis**traction
eu–	good, pleasant	**eu**genics
ex–, ef–, e–	away from, out	**ex**ternal
il–, ir–	not	**il**legal
in–, im–	not, in, into	**in**ject
inter–	between, among	**inter**vene
mal–	bad	**mal**ice
mis–	badly, poorly, not hate	**mis**use **mis**anthrope
post–	after	**post**script
pre–	before	**pre**cede
pro–	before; in place of	**pro**ceed
re–	again, back	**re**submit
semi–	half	**semi**circle
sub–, suf–, sum–, sus–, sup–	under, beneath	**sub**marine **sus**pended
syn–, sym–, syl–, sys–	together, with	**sym**pathy **syn**onym
tele–	far, distant	**tele**scope
trans–	across, beyond	**trans**late
ultra–	beyond, excessively	**ultra**modern
un–	not, opposing	**un**true

These suffixes occur frequently in English words:

Suffix	*Meaning*	*Example*
–able, –ible	able, likely	vis**ible**
–an, –ian	one belonging to, pertaining to	Mexic**an** Mart**ian**
–ance, –ancy, –ence, –ency	act of, condition, fact	veng**eance** presid**ency**

Suffix	*Meaning*	*Example*
–ant, –ent	actor, agent, showing	ten**ant**
–dom	state of, rank, condition	duke**dom**
–er, –or, –eer	doer, maker, dealer in, worker at, one who	lectur**er** engin**eer**
–fy, –ify	make, cause to have	cod**ify**
–ic	dealing with, caused by, person or thing showing	automat**ic** psych**ic** fanat**ic**
–ist	doer, believer	public**ist**
–ize	make, cause to be	demoral**ize**
–ous, –ose	marked by, given to	fam**ous**

practice 1

Without using the dictionary, list five words that start with each of the prefixes given. Check the list in the lesson for prefix definitions.

1. bi–
2. dis–
3. ex–
4. in–
5. inter–
6. mis–
7. de–
8. pre–
9. trans–
10. post–
11. sub–
12. un–
13. pro–
14. tele–
15. sym–

practice 2

Using the prefix list in the lesson, find the correct prefix for each definition given in parentheses. Then write the word that results when the prefix is added to its base.

1. (again) + **play**
2. (away from) + **pulsion**
3. (down) + **tain**
4. (hate) + **anthropic**
5. (after) + **script**
6. (against) + **dict**
7. (away from) + **plode**
8. (half) + **annual**
9. (far) + **vision**
10. (excessively) + **liberal**
11. (not) + **fair**
12. (not) + **visible**
13. (two) + **weekly**
14. (good) + **thanasia**
15. (lacking) + **symmetrical**
16. (not) + **conclusive**
17. (against) + **biotics**
18. (together) + **onym**
19. (around) + **navigate**
20. (across) + **mit**

practice 3

Using the suffix list in the lesson, find the correct suffix for each definition given in parentheses. Then write the word that results when the suffix is added to the word in heavy type. Check the dictionary for spellings.

1. the **wise** (state of)
2. **overseeing** (worker at)
3. **transferring** (act of)
4. **fertile** (cause to be)
5. **pores** (marked by)
6. **supervises** (one who)
7. to be **noticed** (able)
8. **Darwin** (pertaining to)
9. **Mexico** (one belonging to)
10. **notice** (cause to have)

← application →

Read the following paragraph and write the words in heavy type on your paper. Divide each word into its parts. Under each part, write its definition as given in the lesson. If a word doesn't make sense, consult a dictionary and write the dictionary definition on your paper.

> Old Dawson cast a **misanthropic** glance at the new **scribe.** A **psychologist** would have said that Dawson felt the **malevolence** of those approaching retirement when they are forced to **credit** the young of the world with more than simple **logic.** The fact was that young Marcus seemed to be a most **judicious** fellow, **cognizant** of his place and yet unwilling to bow or scrape because of his lowly job in the firm. He had already suggested a stock purchase to his, and Dawson's, boss that had got Marcus the reputation of an **archpriest** with a direct line to the gods of Bay Street. There's no telling, Dawson thought, what a **sophist** like Marcus might come to.

65 The Dictionary and Word Parts

Breaking down an unfamiliar word into familiar word parts can sometimes help you guess its meaning. If you recognize the prefix *un–*, the base *eat,* and the suffix *–able,* the meaning of a word like *uneatable* is easy to figure out. It must mean "not able to be eaten."

But deciding the meaning of a word by adding up the meanings of its parts doesn't always work so easily. For instance, analyzing word parts won't work if you can't recognize what the parts are or don't know what they mean. Suppose you don't know the meaning of *preclusion* in this sentence.

> An investigation of the candidate's involvement in the bank scandal led to his **preclusion** from the ballot.

Even if you recognize *pre–* as a prefix meaning "before," that's not much help if you don't know what the base *–clus–* means or if you forget that *–ion* is a suffix. Since trying to break *preclusion* into parts hasn't helped you determine its meaning, there is only one thing to do. Look the word up in the dictionary. When you do, you'll find that you'll still need an understanding of word parts and their meanings. Here's why.

If, as is likely, you're using a desk dictionary, you won't find *preclusion* listed as a main entry by itself. Instead, you'll probably find it listed as part of another entry, like this:

> **preclude** (pri-klōōd′) *vt* **–cluded; –cluding. 1.** To make impossible by previous action; to prevent. **2.** To bar (a person) from something. — **preclusion** *n* — **preclusive** *adj* — **preclusively** *adv*

The noun *preclusion* appears at the end of the entry for a word built from the same prefix and base—the verb *preclude.* Although the dictionary has given you the meaning for *preclude,* you'll have to determine the meaning of *preclusion* yourself. Here's how.

First, look up the meaning of the suffix *–ion.* Dictionaries give meanings for prefixes and suffixes as well as for words.

> **–ion** *n suffix* **1.** An act or process, or the outcome of an act or process. **2.** A state or condition

Then, combine the meaning of *preclude* and the meaning of *–ion* that best fit the sentence in which *preclusion* appears. In this way you can determine that, in this sentence, *preclusion* must mean "the act of barring a person from something."

Remember that even if you use the dictionary to find the meaning of a word, you still need to understand how words divide into prefixes, bases, and suffixes.

practice 1

Here is a list of ten words followed first by definitions of suffixes and prefixes and then by definitions of words closely related to the list words. Use these definitions to determine the meanings of the ten list words. Then write each list word and its meaning.

1. disparagement
2. cataclysmic
3. ubiquitousness
4. extraneously
5. reescalation
6. captivator
7. nonshrinkable
8. perpetuation
9. ineptitude
10. chronologically

–or, –er	one who	**–ic**	tending to produce, caused by
–able	capable of	**–ly**	in a specified manner
–ence **–ation** **–ion**	act or process, state or condition	**–ment**	an action or result of an action
–ness **–tude**	state or condition, quality	**non–**	not
		re–	again

captivate to influence by charm, art, or great appeal—**captivator**

chronological arranged according to the order of time—**chronologically**

ubiquitous existing everywhere at the same time; widespread—**ubiquitousness**

cataclysm a violent change; a momentous event marked by upheaval and destruction—**cataclysmic**

disparage to degrade or belittle; to speak slightingly about—**disparagement**

extraneous coming from the outside; not belonging; not relevant—**extraneously**

inept unfit, inappropriate, incompetent—**ineptitude**

perpetuate to cause to last indefinitely—**perpetuation**

shrink to contract; to become smaller—**shrinkable**

escalate to increase; to expand—**escalation**

practice 2

Look up each of the following words in the dictionary. If you can't find a word listed as a main entry, first look up the meaning of the part of the word printed in heavy type. Then look up the meanings of the other parts of the word. Combine these meanings to arrive at a definition for the whole word. Then write the word and its definition.

1. **deferential**ly
2. re**investig**ative
3. **ostentatious**ness
4. non**biodegradable**
5. **inexorabil**ity
6. un**inhabit**able
7. **corroborat**ion
8. un**statesman**like
9. **rigorous**ness
10. **depreciat**ion

← application →

Each of the following words is closely related to a word you worked with in practice 1 or practice 2. Using what you learned in these practices, write a short definition for each word. Then check your answers in the dictionary.

1. ubiquitously
2. ineptly
3. perpetuator
4. extraneousness
5. inexorably
6. captivation
7. ostentatiously
8. investigation
9. rigorously
10. disparager

INDEX